Technical Standards and Design Guidelines

Mixed - Use Buildings

RANJIT GUNEWARDANE

AuthorHouse™
1663 Liberty Drive
Bloomington, IN 47403
www.authorhouse.com
Phone: 1 (800) 839-8640

Published by AuthorHouse 08/09/2018

ISBN: 978-1-5462-4327-4 (sc)
ISBN: 978-1-5462-4326-7 (e)

Library of Congress Control Number: 2018906375

Print information available on the last page.

This book is printed on acid-free paper.

authorHOUSE®

Commercial Building Technical Standards and Design Guidelines

Technical Standards and Design Guidelines

MIXED - USE BUILDINGS

Table of Contents

1.0 Introduction, Purpose and Intent xi

1.1. Introduction/Purpose 1
1.2.0 Design Philosophy 2
1.3.0 Mixed use Concept Plan 3
1.4.0 Using the Design Guidelines 3
1.5.0 General Requirements, Reviews and Approvals .. 3
1.6.0 Variances 4
1.7.0 Code Compliance 4
1.7.1 Introduction 4
1.7.2 Occupancy Classification 5

2.0 Architectural Design Criteria 7

2.1 Architectural Imagery 8
2.1.1 Introduction 8
2.1.2 Massing/Orientation 9
2.1.3 Building Façade 9
2.1.4 Exterior Illumination 10
2.1.5 Signage 11
2.1.6 Entry 11
2.1.7 Parking 12
2.1.8 Landscaping 13
2.1.9 Site Utilities and Mechanical Equipment 13
2.2.0 Loading Dock and Trash Storage Areas 14
2.3.0 Life Safety 17
2.3.1 Overview 17
2.3.3 Fire/Smoke Separations 17
2.3.4 Fire Fighting Access 18
2.3.5 Means of Egress 18
2.3.6 Fire-Rated Doors 19
2.3.7 Ducts and Air Transfer Openings 20
2.3.8 Finishes 20
2.3.9 Miscellaneous 20

3.0 Building Site 21

3.1.1 Site Survey 22
3.1.2 Orientation 22
3.1.3 Wind Direction 23
3.1.4 Topography 23
3.1.5 Building Placement 23
3.1.6 Traffic Access General 24
3.1.7 Parking & Garages 26
3.1.8 Landscape 27
3.1.9 Security and Control 27
3.2.0 Plumbing systems 28
3.2.1 Site Amenities 28

4.0 Facility Description and Functional Relationships 30

4.1.1 Space Standards 31
4.1.2 Accessibility 31
4.1.3 Environmental and Architectural Parameters 32
4.1.4 Criteria Matrix 32
4.1.5 Planning Brief - Commercial Areas 33
4.1.6 Retail Shop Design 34
4.1.7 Restaurant/Food Tenants 35
4.1.8 Conference/Meeting Facilities 37
4.1.9 Back of House Support Areas 41

4.2.1 Health and Fitness Suites ... 43
4.2.2 Swimming Pools ... 49
4.2.3 Outdoor Tennis Courts ... 51
4.3.0 Commercial Offices ... 51
4.3.1 Space Plan Components ... 52
4.3.2 Work Area Mix Criteria Matrix ... 52
4.4.0 Residential Apartments ... 54
4.4.1 General ... 54
4.4.2 Space Plan Components ... 55
4.4.3 Space Planning Concepts ... 56
4.5.0 Back of House Areas ... 61
4.5.1 Storage ... 61
4.5.2 Housekeeping Department ... 62
4.5.3 Laundry ... 62
4.5.4 Employee Facilities ... 62
4.5.5 Maintenance and Engineering ... 63
4.5.6 Back of House Corridors ... 63
4.5.7 Receiving Area/ Loading Dock ... 63
4.6.0 Major Equipment Spaces ... 63
4.6.1 Chilled Water Plant ... 64
4.6.2 Fan Rooms ... 64
4.6.3 Domestic Water Pumps ... 64
4.6.4 Electrical Service Entrance, Transformers, Switchgear, and Emergency Power Supply ... 65
4.6.5 Elevators (Lifts) ... 65
4.6.6 Service Cores ... 65
4.6.7 Swimming Pool Plant Rooms ... 66

5.0 Building Materials, Construction and Finishes ... 69

5.1.0 Building Materials and Design ... 70
5.1.1 General ... 70
5.2.0 Building Exterior and Entryway Design ... 73
5.2.1 General ... 73
5.2.2 Entryways ... 73
5.2.3 Materials, Construction and Finishes ... 74
5.3.0 The Main Entrances and Concourse ... 76
5.3.1 General ... 76
5.3.2 Materials, Construction and Finishes ... 76
5.4.0 Retail Stores ... 78
5.4.1 General ... 78
5.4.2 Materials, Construction and Finishes Storefront:79
5.5.0 Full-Service Food Court Criteria ... 82
5.5.1 General ... 82
5.5.2 Materials, Construction and Finishes ... 82
5.6.0 Public Lift Lobbies ... 84
5.6.1 Program Requirements ... 84
5.6.2 Materials, Construction and Finishes ... 85
5.6.3 Public Lift Cabs ... 86
5.6.4 Materials, Construction and Finishes ... 86
5.7.0 Public Washrooms ... 87
5.7.1 Facilities Include Public Washrooms for Handicapped Visitors ... 87
5.7.2 Materials, Construction and Finishes ... 88
5.8.0 Prayer Rooms ... 90
5.8.1 Facilities Include: ... 90
5.8.2 Materials, Construction and Finishes ... 90
5.8.3 Ablution Area ... 91
5.8.4 Materials, Construction and Finishes ... 91
5.9.0 Security Control Room ... 92

6.0 Recreational Facilities ... 95

6.1.0 Swimming Pools ... 96
6.1.1 General ... 96
6.1.2 Materials, Construction and Finishes ... 97
6.1.3 Pool Snack Bar ... 99
6.1.4 Pool Locker Rooms ... 100
6.1.5 Special Requirements ... 101
6.2.0 SPA ... 101

6.2.1 General..101
6.2.2 Multi – Purpose Therapy Rooms 101
6.2.3 Relaxation Lounge 102
6.2.4 Materials, Construction and Finishes 102

7.0 Commercial Offices 106

7.1.1 General 107
7.1.2 Interior Architecture 107
7.1.3 Materials, Construction and Finishes 112

8.0 Maintenance and Engineering 116

8.1.1 General 117
8.1.2 Materials, Construction and Finishes 117
8.1.3 Mechanical/ Electrical Room 118
8.1.4 Materials, Construction and Finishes 119

9.0 Engineering Design Criteria.................................... 121

9.1.1 General 122
9.1.2 Design Standards 122
9.1.3 Designers Responsibilities.................................123
9.1.4 Technical Submissions 123
9.1.5 General Design / Installation Principles 124
9.1.6 Deviations from the Owner's Standards 124
9.2.1 Utility Services...125
9.3.1 Life Safety Systems...126
9.3.2 Fire Protection Systems 126
9.3.3 Fire Alarm and Detection..................................127
9.4.0 Fire Suppression 131
9.4.1 Automatic Sprinkler System...............................131
9.4.2 Standpipe Systems 132
9.4.3 Clean Agent Extinguishing Systems...................132
9.4.4 Cooking Equipment Suppression Systems.......132
9.4.5 Portable Fire Extinguishers 134
9.5.0 Emergency Lighting 134
9.5.1 Objectives 134
9.5.2 Design Standards 134
9.5.3 Control Requirements.......................................135
9.5.4 Description..135
9.6.0 Smoke Control/Clearance..................................135
9.6.1 General..135
9.6.2 Smoke Control 136
9.6.3 Stairway Pressurization.....................................136
9.6.4 Control systems..136
9.6.5 Testing 137
9.7 LPG Storage 137
9.7.1 General 137
9.7.2 Storage Tanks...137
9.7.3 Emergency Isolation and Detection.....................138

10.0 Mechanical/HVAC 139

10.0 Mechanical/HVAC – All Areas.............................140
10.1.1 General 140
10.1.2 Systems..141
10.1.3 Design Criteria..141
10.1.4 Aims and Objectives....................................... 143
10.1.5 Thermal Zoning 144
10.1.6 Heat Recovery and Energy Conservation Objectives 148
10.1.7 Source Components: Cooling 148
10.1.8 Control Requirements: 149
10.2.1 Building Management System (BMS) 150
10.2.2 Wired and Wireless Technologies 150
10.3.1 Chilled Water Production/Distribution 152
10.4.1 HVAC System Selection 154

11.0 Electrical Installation 157

11.1.1 General 158
11.1.2 Power/Electrical...158

11.1.3 Electric Power Management system 159
11.1.4 Energy Efficiency 161
11.1.5 Emergency standby power to dedicated supplies 161
11.1.6 Uninterruptible Power Supply (UPS) 165
11.1.7 Local Distribution/Protection 165
11.1.8 Grounding and Bonding 168
11.1.9 Motor Control Center (MCC) 168
11.2.1 Lightning Protection 169
11.3.1 Lighting 169
11.4.1 Security Systems 175
11.4.2 Integrated security systems 175
11.5.1 Energy and Water Metering 180
11.5.2 Sub-Metering 181
11.5.3 Meter/Sub-Meter Performance Metrics and Attributes 182
11.5.4 Communication Networks and Data Storage Requirements 182

12.0 Plumbing and Sanitary Systems 184

12.1.1 General 185
12.1.2 Domestic Hot and Cold Water 185
12.1.3 Sanitary Drainage 186
12.1.4 Storm Water Drainage 187
12.1.5 Water Treatment 187
12.1.6 Gas Supplies 188

13.0 Vertical Transportation 189

13.1.1 General 190
13.1.2 Elevators/Lifts. 190

14.0 Audo/Visual (A/V) Systems 194

14.1.1 General 195
14.1.2 Coordination: 195
14.1.3 Infrastructure Cable Plant Design 197
14.1.4 Audio/Video Systems 198
14.1.5 Function Rooms and Boardroom 202
14.1.6 Ballroom 203
14.1.7 Food & Beverage Outlets 204
14.1.8 Staff Training Room 204
14.1.9 Video Conferencing System 205
14.2.1 Digital signage 206
14.3.1 Networked A/V Security 206
14.3.2 AV / IT Security Framework 206

15.0 Technology 208

15.1.1 Technology Statement of Direction 209
15.1.2 Physical Environment 209
15.1.3 Security 210
15.1.4 Electrical 210
15.1.5 Disaster Prevention 211
15.1.6 Environmental 212
15.1.0 Computer Room (Data Centre) 213
15.2.1 PBX Room 215
15.3.1 IP TV/ Head End Room 215
15.4.1 High Speed Internet Access 215
15.5.1 Main Distribution Frame 216
15.6.1 Uninterruptible Power Supply 216
15.6.2 To determine the size of UPS required; 217
15.7.1 Cabling System Standards 218
15.7.2 Cabling Infrastructure 219
15.8.1 Intermediate Distribution Frames 219
15.9.1 Local Area Network Standards 220
15.9.2 Core Layer 221
15.9.3 Distribution Layer 222
15.9.5 Network Security 222
15.9.6 Firewall, Antivirus, Malware 223
15.9.7 Network Management 223
15.9.8 Hardware 224

15.9.9 Switches, Routers, Peripherals........................ 224
15.9.10 Servers, User Devices - Hardware Standards 225
15.10.1 Telecommunication Standards........................ 226
15.10.2 Telephone (iP PBX) System 226
15.10.3 Voice Mail (VM).. 228
15.10.4 Call Accounting.. 229
15.10.5 End User Devices.. 230
15.11.1 WI-FI 230
15.11.2 Design Overview – WIFI and Fixed Wire......... 230
15.11.3 General Standards 231
15.11.4 Wireless Access Pointsk................................. 232
15.11.5 HSIA Controllers.. 233
15.11.6 Service Management Platform 233
15.11.7 Bandwidth Management.................................. 234
15.11.8 Security 234
15.12.1 Internet Support 234
15.13.1 Building Internet of Things 235

16.0 Testing & Commissioning 239
16.1.1 Factory Testing (Equipment)............................. 240
16.1.2 Site Testing 240
16.1.3 Commissioning 240
16.1.4 Method Statements...241
16.1.5 Program..241
16.1.6 Record Documentation 245
16.1.6.1 Record Drawings... 245
16.1.6.2 Operating and Maintenance Manuals 246
Appendix 1 – Acoustical Performance........................ 260
Appendix 1I – Criteria Matrix...................................... 262

17.0 Glossary.. 263

SECTION 1.0

Introduction, Purpose and Intent

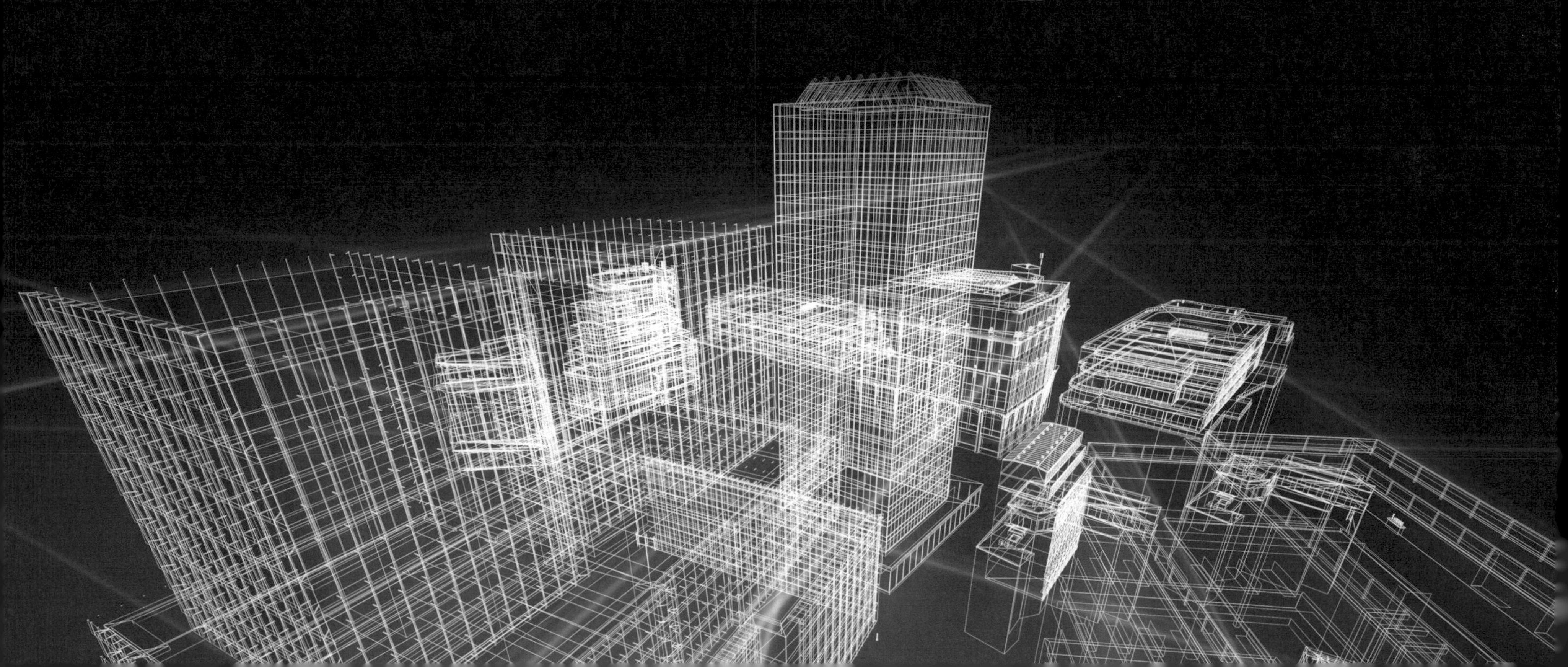

1.1. Introduction/Purpose

A Mixed Use (MU) Development is defined as “Development that integrates two or more land uses, such as residential, commercial, and office, with a strong pedestrian orientation. The purpose of the Mixed-Use Development Technical Standards & Design Guidelines (TSDG) Manual is twofold. First, it is intended to serve as a central reference for all zoning and subdivision code requirements relating to the mixed-use zone districts and street standards for mixed use development in the proposed building location. Second, since mixed use development differs in significant ways from the design of the prevailing single-use development, the Mixed-Use Development TSDG is meant to provide a guide to the application of the design standards for mixed use through illustrations and design guidelines. This document contains scoping and technical requirements for the Architectural, Engineering systems, and Technology applications design.

The intent is not to create a standard design but rather one of quality and consistency. It will establish a defined level of client expectation and a means to measure and maintain a basic level of quality. These requirements are to be applied during the design, construction, addition to, and alteration of sites, facilities, buildings, and elements to the extent required by regulations issued by local authorities.

Standards are intended to be minimum requirements used by Architects, and other specialist design professionals as a guide in the design process. When specific design issues are not addressed, these Standards will be used as a guide to establish design intent in order to develop innovative design solutions which meet or, as governed by market forces, exceed the intent of the Standards. The Design and Construction Standards are divided into the following six (6) sections:

I. Introduction, Purpose and Intent

This section describes general procedures for compliance with Design and Construction Standards. Design should conform to the latest version of the International Building Code, and to Qatar National Construction codes and standards where available, and more stringent.

II. Architectural Design Criteria

This section outlines the requirements of the overall building for image, orientation, materials, life safety, etc.

III. Facilities Program and Functional Relationships

This section describes the space planning,general finishes, construction, power, lighting and special requirements for all functional uses of the mixed-use building.

IV. Engineering Design Criteria

This section describes the minimum performance criteria required for HVAC, electrical, plumbing, fire protection and vertical transportation systems.

V. Audio Visual and Multi-Media Systems

This section will provide a guideline to the professional engaged in Audio-Visual design disciplines. The intent is to establish minimum and sufficient technical criteria, and typical functional requirements.VI. Information & Communications, Technology

This section describes universal solutions capable of supporting all communications needs of the facility that support voice, data, LAN’s etc.

1.2.0 Design Philosophy

Architectural design of the highest quality and being appropriate in their respective settings is essential to the success of the Mixed-Use (MU) building. It is the intent of the Owner to create a design that respect local building methods, new technologies and materials, as well as local cultural, religious and economic factors.

The design of a new building shall be in reference to its site and the character of its surroundings.

Additions shall be respectful of the existing building. Alterations and renovations of existing buildings shall enhance the building features and be respectful of the original building’s contextual surroundings. The “style” of the building should result from local indigenous architectural concepts being re-defined and re-invented. The resulting style should fit seamlessly and harmoniously into its setting, whether urban or suburban in nature.

It is essential that indigenous building methods that have qualities that are environmentally sensitive, and come from renewable source material, and allow for reduced energy consumption are used. Designers should be required to create sustainable designs to international standards, and to optimize the whole-life costs of facilities. The environmental impact of the materials and processes used in the construction of projects should be considered. The design should give the building personality and identity for easy recognition from a distance, with identifying features in the way of signage, canopy etc. that can be read by pedestrians. The entrances being the first opportunity to establish a positive impression and a strong sense of arrival with visitors should be easily identifiable, well landscaped and detailed and well-lit at night to create an inviting and safe environment.

The building shall be designed to optimize rentable area ratios and allow for multiple business types to occupy the facilities over the life of the building.

The main entrance lobby should make bold statements and lasting impressions, as the lobby quickly sets the MU building’s tone and ambience. Therefore, the planning of the major lobby elements and the design of the details are crucial.

The design should balance Visual impact and Function. The planning and design of the administration offices and other back-of-house or service areas of the MU complex, most of which the visitors rarely see, are equally critical to the eventual success of the property operation.

These Technical Standards and Design Guidelines are written as minimum standards for both new construction, renovation, and conversions. It will generally be easier to meet these Standards with new construction without the constraints of an existing structure inherent in renovations and conversions. However, renovation of existing structures should adhere as closely as possible to the standards outlined herein. When meeting, such standards is physically impossible or economically impractical, then alternate methods which achieve the same or similar level of quality should be used.

1.3.0 Mixed use Concept Plan

The Architect shall provide a MU concept plan that describes and illustrates, in written and graphic format, the intended locations and quantities of proposed uses, the layout of proposed vehicle and pedestrian access and circulation systems, and areas designated to meet requirements for open space, parking, on-site amenities, utilities and landscaping. It shall include statements or conceptual plans describing how signage and lighting will be designed in a unified and integrated manner on site. In addition, the MU concept plan shall indicate how the proposed uses will relate to the surrounding properties.

1.4.0 Using the Design Guidelines

The primary use of the Mixed-Use Development TSDG Manual is as a reference document that can be utilized and applied to any phase of the planning, development, and review of a MU project.

This TSDG document will assist architects and other design professionals in developing imaginative, efficient and marketable design solutions as well as ensure a safe and environmentally sound structure, while developing plans for new construction, renovation of existing commercial building facilities and conversions. The TSDG requires that higher standards be met beyond local codes.

Therefore, it is important for the Owner to engage the services of competent design professionals who are familiar with these requirements and how they affect mixed use building design.This will serve to speed up the design as well as approval process of local governing authorities. It will also save time and money through the development of creative solutions which comply with legislation.

This policy recognizes the principal of consistency in maintaining a higher quality at all properties, regardless of the extra expenses or lower standards allowed by local standards. These provisions are in addition to the regulations contained and described in the following sections.Where the provisions of this manual conflict with provisions in the Qatar National Construction Codes, the higher standard should apply.

The terms "shall" and "must" are used where practice is sufficiently standardized to permit specific delineation of requirements, or where safeguarding of the public health and safety justifies such definite action. Other terms, such as "should," "recommended," and "preferred," indicate desirable procedures or methods, with deviations subject to individual consideration.

1.5.0 General Requirements, Reviews and Approvals

Design is a creative activity by which client's needs and objectives are collected, interpreted and expressed in three-dimensional physical solutions. A Design Brief 'Brief to Architect' drawn up by the Owner appointed specialist consultant, should clearly list down and describe in a simple & easy to understand manner, all the requirements & expectations from the design, and given to the Architect and his team, as this is the first & most critical touch point in the MU building Design process. This 'Brief to the Architect' shall be properly documented & presented to the architect who will work on the project. The intent of this document shall be to provide an in depth understanding of the Owner's vision for the Multi-Use building in terms of design expectations & requirements from a retail and operational perspective, which would in turn assist the architect in designing the building complex in tandem with the Owner's vision/ findings from a market research etc. -

This 'Brief to Architect' is intended to assist the architects, and specialist consultants to create the project in the manner conceptualized & agreed, also ensure that the design is in harmony with the market demand & the consumer expectations. Design management is an important activity in the design process as it involves the coordination,analysis and testing of the design, as well as the management of the different stakeholders involved. To ensure that optimal design, value for money and buildability are achieved, due care and attention need to be given to the proper management and coordination of all design activities throughout the design process. The interaction between the different design disciplines requires a well-coordinated teamwork structure. Design management encompasses all the coordination, analysis and design testing activities that a project requires. For effective design management and coordination, it is necessary to appoint a manager with appropriate management skills to ensure the design process operates efficiently. Such a person is usually the design team leader. To ensure its effectiveness, the Project Coordinator should draw up a program which includes the main areas of activity (i.e. Planning, Implementation and Review) up to project occupation. In relation to the design development activity (part of the Planning Stage) the Design Team Leader should, as soon as the other principal consultants are appointed, draw up details of design responsibilities and milestones for each consultant and illustrate them on a project program which should be part of the Project Execution Plan. Design should be a staged process during which several approvals / sign-offs are required from the client. Approvals are usually given as part of the formal project review structures. In each case, there should be sufficient information for the client to give informed approval. The timing and sequencing of client approvals may differ from project to project, depending on how the design process is carried out. Various regulations and laws apply to the design of projects, both building and civil engineering, and to their owners and users. Projects should be designed so that approvals from all relevant Statutory Authorities can be obtained. The Owner's representative should review designs regularly to ensure that they satisfy the needs expressed in the Definitive Project Brief and Sign off on designs.

1.6.0 Variances

The purpose of these Design Guidelines is to establish a level and consistency of quality in design and construction of mixed-use commercial properties. However, due to variations in building construction, site conditions and other variables, some aspects of these guidelines may not be possible to achieve. When specific requirements cannot be met, alternative solutions should be explored which achieve the same or similar level of quality and meet the intent of the guidelines.

When considering variances, a distinction should be made between new construction and the renovation or conversion of existing structures. Renovation of existing structures will on occasion, contain physical constraints which make full compliance impossible or economically feasible.

1.7.0 Code Compliance

1.7.1 Introduction

This design objective shall maintain harmony yet produce diversity so that visual interest may be achieved through the adoption of specific building codes and regulations controlling the configuration, features, and functions of the building that define and shape the public realm.

The following is an outline of the requirements at the stage of the design. In addition to the latest version of the International Building Code (IBC), the full family of International Codes published by the International Code Council (ICC) and ADA Standards for Accessible Design should be applied to the following aspects of the design:

Fire and life safety features, Architectural space planning (including Accessible Design) Building classification and fire separation requirements. Mechanical Systems, Plumbing Systems, Fuel Gas Appliances Qatar Electricity and Water Corporation (KAHRAMAA) Standards Other codes and standards prevailing in the State of Qatar should be applied on components of the building if they are higher than described in the TSDG. The Architect shall assume full responsibility for the compliance with all applicable International and local regulations pertaining to design work.

1.7.2 Occupancy Classification

The occupancy classification under the IBC for the various uses in the MU Building project should be as follows:
Occupancy Group Description A – 2 Assembly Uses intended for food and Beverage Consumption

- A – 3 Assembly Uses intended for recreation or amusement
- B – Offices, Banks and Storage of Records and Accounts
- M – Mercantile – Retail
- R – 2 Residential
- S – 1 Storage of Moderate Hazard Items
- S – 2 Parking Garages

The covered parking garage floors should be mechanically ventilated and should not qualify for an open parking garage. Based on the above, the following requirements for the project should be verified during design development.

When two or more Occupancy Groups are combined in one building, the IBC allows these mixed uses to be treated as either Non-separated or Separated Occupancies.

When occupancies are Non-separated, within each occupancy area, occupant load calculations, egress configuration, and other such requirements shall be applied per the code restrictions for that occupancy.

Separated Occupancies shall be segregated from one another by fire separations, which may consist of fire-resistant walls, fire doors and other rated openings, and fire-resistant floor/ceiling assemblies. The degree of fire resistance required for such separations will vary from 1 to 4 hours, depending on the occupancies involved and whether the building is sprinklered.

Other uses may, at the designer’s option, be treated as Accessory or Incidental to the major occupancy within which they occur, rather than per the Separated or Non-Separated Occupancy provisions. Incidental Uses areas consist of a specific list of uses which may be treated in a manner like Accessory spaces. However, because of their higher degree of hazard, additional protection shall be required in the form of fire-extinguishing systems, fire sprinklers, and/ or rated fire separations between the Incidental Use and the major occupancy within which it is located. Except for these protections, all other requirements, such as height and area requirements, shall be governed by the major occupancy.

Examples of Incidental Uses and their protection requirements include the following:

Boiler rooms with equipment over certain size limits, with either a one-hour rated separation or an automatic fire- extinguishing system and construction capable or bc Where areas are not provided with fixed seating, the occupant load shall be based on a calculation using the floor area of a space. In spaces having fixed seating, the occupant load shall be based on the number of seats.

In Mixed-Use buildings, the design of the means of egress shall apply to each portion of the building based on the use of that space. The number of exits or exit access doorways from spaces shall be as outlined in section 1015/1016.1 of the IBC. The travel distance to an exit from the dead end of a corridor shall not exceed half the distance specified.

The distance between the exit doors should comply with the separation of exits provisions of Section 1015 of the IBC. All exit enclosures that serve four or more stories should be of 2-hour fire resistive construction as per Section 1022.2. The access to exit corridors at each floor should be separated from the occupied space by a 1-hour fire resistive construction complying with the requirements of Section 1018 of the IBC. The two dead-end portions of a corridor shall not be more than 6 M, thus complying with the provisions of Section 1018.4. Section 403.5.4 of the IBC requires the exit enclosures to be pressurized in accordance with Section 909.20 and 1022.10 where the height of the building is greater than 23 m.

SECTION 2.0

Architectural Design Criteria

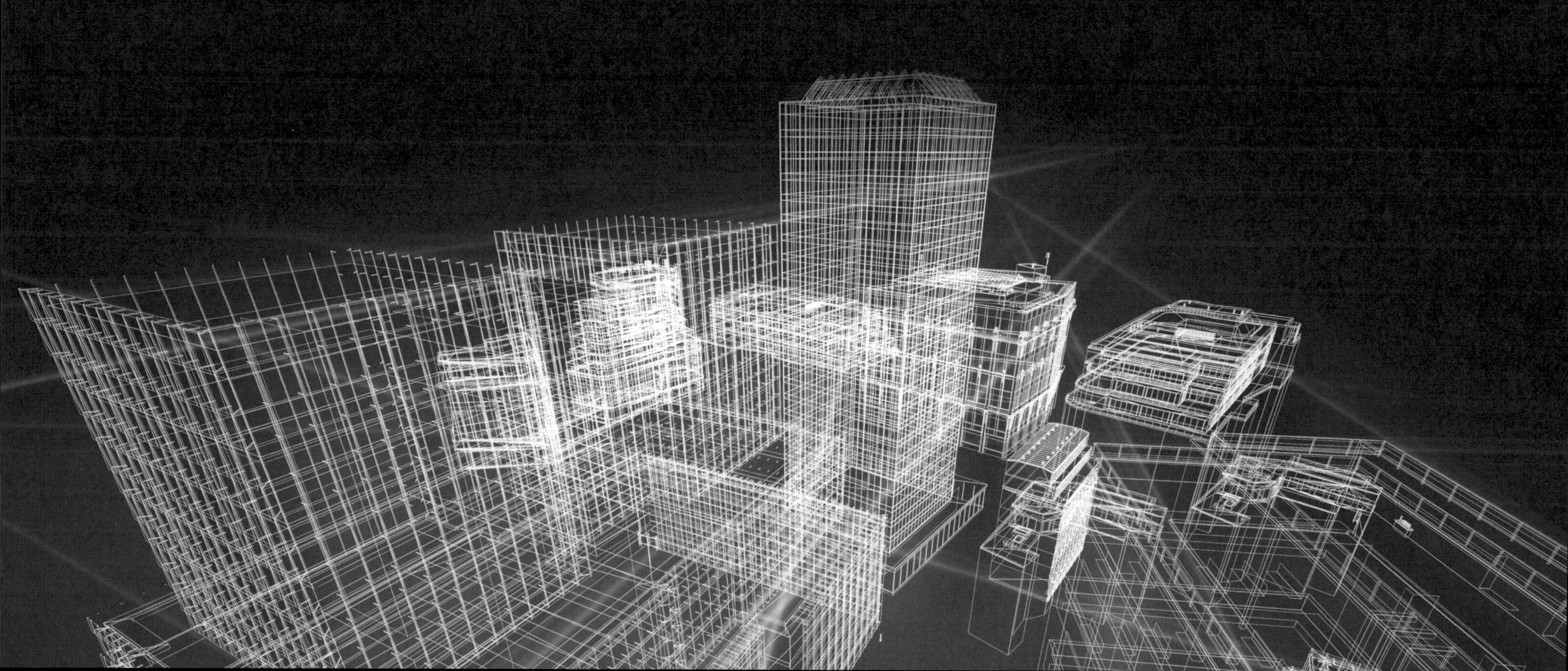

2.1 Architectural Imagery

2.1.1 Introduction

Large retail developments depend on high visibility from major public streets. In turn, their design determines much of the character and attractiveness of major streetscapes in the location. Major cities and suburban towns have a development review system that promotes solutions to general issues. The purpose of these technical standards and design guidelines is to augment those existing criteria with more specific interpretations that apply to the design of large Mixed-Use developments. These standards and guidelines require a basic level of architectural variety, compatible scale, pedestrian and vehicle access, and mitigation of negative impacts.

The standards are by no means intended to limit creativity; it is the Client's expectation that they will serve as a useful tool for design professionals engaged in site specific design in context. A new MU development can make an important contribution to the vitality and vibrancy of city and town centers, providing greater shopping choice for the consumer, and increased activity in local economies. However, successful outcomes also demand that a high quality of design is employed to enhance the 'sense of place' of the location.

The key to the successful integration of a new development into city and town centers locations is regard for the basic tenet of urban design that each building is part of a greater whole, and whatever the merits of any individual development, its contribution to improving the overall character and quality of its location is a key consideration. While built form, scale and mass should have regard for its urban context, this shall not infer that new development has to replicate local building traditions or mimic adjacent structures; on the contrary, new development should express its function in an architecture that is of today, but nevertheless with regard for the topography and morphology of its location.

New MU development should be of a high design standard and wherever generic building types are proposed, their designs should be adapted to ensure that they contribute positively to the character and quality of the location. Building elevations must consider the character of the surrounding architecture and neighborhood and incorporate design elements to further enhance community character. Three hundred and sixty degrees (360°) architecture is generally required. All sides of all buildings are to be treated with the same architectural style, use of materials, and details as the front elevation of the building. The lower stories of the building referred to as the base building shall frame the public realm, articulate entrances, and assist in the creation of an attractive and animated public realm which provides a safe, interesting, and comfortable pedestrian experience. The base building shall and support adjacent streets, and open space at an appropriate scale, assist to achieve transition down to lower scale buildings, and minimize impact on parking and servicing on the public realm.

A single building or development of multiple buildings within a development must maintain a consistent style/ architectural theme. Architectural design, building materials, colors, forms, roof style and detailing should all work together to express a harmonious and consistent design.This includes all "pads" within a retail development or other accessory structures. Where a large floor plate is required to accommodate the needs of a retail sector, and the scale and mass of development is greater than adjacent buildings, creative design solutions - and flexibility in the operations of the client - can successfully adapt generic design templates to the urban structure, and the character and context, of the location.

2.1.2 Massing/Orientation

In MU centers the development of a complex of buildings is preferable to a single large structure so that the varied massing provides visual interest and human scale. Additionally, the spaces created between the various buildings will provide opportunities for pedestrian plazas, courtyards and other outdoor gathering areas.

The Architect shall use solar resource information for the building location, to assist design passive solar and day lighting features, and optimize the building shape. Buildings shall be designed to provide protection from prevailing winds, and to enhance public views of features, and to minimize obstruction of views from adjoining structures. The designer shall utilize human scaled massing, varied articulation treatments, and traditional facades.

Large MU projects with street frontages greater than 30 m shall incorporate traditional massing and facade techniques such as: Dividing the facade into modular bays a minimum of every 7.5 m; Buildings shall utilize elements such as massing, materials, windows, canopies and articulated roof forms to create a visually distinct base as well as a cap. Using traditional architectural detailing (i.e. ornamentation, window placement, changes in materials and/or colors) as opportunities to bring a human scale to a larger frontage.

The scale of all structures in relationship to other structures and spaces is important. Buildings and the spaces between should relate easily and openly to the external public areas. To balance horizontal features on longer facades, vertical building elements, such as building entries, should be emphasized. Shadow Studies shall be modelled by using commercial software and ensure that the building will not violate the city zoning codes. Minimize sun and wind impacts, and protect sunlight and sky view, for streets, parks, public private open space, and neighboring properties; The impact of development in terms of sun and daylight access to the surrounding context including surrounding buildings, the public realm, public and private open space shall be evaluated. The building design shall ensure greater sunlight penetration on the sidewalk across the street.

The building orientation shall be such that solar heat gain is avoided, and the use of volume and surfaces of the building shall be oriented through studies of the air temperature effect for faces of the buildings in different orientations. The building shall be designed to fit harmoniously within the existing context of neighboring building heights at the street and to respect the scale and proportion of adjacent streets, and public or private open space.

2.1.3 Building Façade

The architect shall provide architectural expression and design elements, such as cornice lines, window bays, entrances, canopies, building materials, and fenestration, in a pattern, scale, and proportion that relate to neighboring buildings and engages pedestrians.Place building entrances and transparent windows on all façades facing streets, parks, and open space. Design the first 10-12 m of façade in accordance with bird friendly practices. A building façade does many things:

Provides thermal control

- Limits air flow
- Delivers daylighting
- Prevents water entry

- The type of external cladding material used on the building has a great impact on water absorption. Even the best, and most carefully installed cladding systems must manage water that penetrates the exterior skin. Depending on the climate zone, the most appropriate of the following three approaches shall be selected to keep wind- driven rain out of walls:
- Barrier systems
- Drainable assemblies
- Pressure-equalized rain-screens

The specialist designer shall provide detailed description of method to be adopted to stop the water at a single plane, the joints between cladding members must be resistant to the penetration of driven rain over time. Project located in not so severely cold or wet climate, a barrier system may be a very practical solution. Where the project is in a climate that has severe weather conditions, a "perfect" air barrier and a "perfect" vapor barrier is required.

Windows and glazing systems shall be carefully selected to ensure the safety and comfort of all occupants, in addition to maintaining character of the building façade. Consequently,the window design and glazing selection are critical to the performance and appearance of the building envelope. Spectrally selective products applicable to glazing technology that allow for higher transmittance of visible light than that of the heat contributing infrared portion of the spectrum shall be provided.

External shading devices shall be an integral part of the building envelope design to limit solar heat gain and glare. Horizontal shading devices should be placed above windows on south facing walls, and vertical louvres/ shading devices would be effective for east, west and north facing windows. Use continuous insulation where possible on walls and roofs, and account for the thermal bridging of window framing and metal or wood studs in walls. The minimum R-value required for the climate zone for the building envelope shall be established using ASHRAE Standard 90.1 – 2016 (International Climate Zones) as a baseline for minimum insulation R-values. The tables outline R-value standards for roof – insulation above deck, Walls Above Grade, Walls Below Grade, floors and Slab- on-Grade Floors shall be followed.

2.1.4 Exterior Illumination

Lighting should be designed to serve to improve security and way-finding and provide important visibility to commercial sites. Additionally, lighting should be used to create special effects and feelings in the nighttime landscape. Lighting scheme shall be developed to prevent glare which affects approaching motorists and pedestrians, and to minimize light trespassing onto adjoining properties.

The purpose for the lighting is a critical consideration. Lighting for security or surveillance requires a different strategy than designing for a more intimate space.

Moonlighting shall be accomplished by using combinations of light carefully located high up in trees and other low wattage ground level lighting attached to branches and leaves from below. Moonlighting could create very dramatic effects and is especially good for transitions between lighted areas.

Back lighting may be used to feature a tree or shrub or another element with an unusual or visually pleasing silhouette. Silhouette lighting can be achieved by directing lighting to a surface that reflects the light into or on a desired area. The maximum height of lighting poles shall be 7.5 m. Pedestrian areas shall be well-marked and well-lit.Exterior lighting shall be an integral part of the architecture and landscape design. Street lighting shall relate in scale to the pedestrian character of the area. Pedestrian lighting shall be provided with the source light being shielded to reduce glare, thereby encouraging safe access to the property twenty-four hours per day. Overall, lighting and pedestrian zone lighting shall not create glare or light spillage off site or beyond parking lots and streets Up-lighting and accent lighting are encouraged within the landscape areas but shall not be directed toward a public or private street or drive aisle. Well-designed lighting should be provided for the operation of effective perimeter security controls. A photometric plan and outdoor lighting report shall be submitted that addresses all aspects of property illumination, including but not limited to the parking area, building, parking, and signage and address mitigation of negative impacts on adjacent residential uses or residentially zoned properties.

All carriageways within the property curtilage should be adequately illuminated with light standards and fixtures of appropriate design and quality. The Main Entrance carriageway should have an average level of illumination of 160 Lux, with a minimum of 380 Lux under the entrance canopy.

2.1.5 Signage

The visual transfer of business advertising and other public information using external signs shall comply with all applicable sign ordinances. and the following design. All completed signs must have a high quality professional appearance. Sign materials, colors, and shades shall be compatible with the related buildings on the property and shall be limited to high quality construction materials, such as stone, brick, or decorativeblock, finished wood (painted or stained), or finished metal.External lighting of monument signs shall be concealed and ground-mounted. Illumination of signs shall be directed away from all traffic and from all adjoining residential areas. The intensity of the light shall not exceed 15-foot candles at any point on the sign face.

Interior non-illuminated tactile building signs including Braille. Braille-Tac™ one-piece construction sign system utilizing an aqueous developing process to produce raised numbers and letters with corresponding Grade II Braille (complying with Specification #800), and pictograms on photopolymer sign, all complying with ADA and CABO/ ANSI A117.1 requirements.

Sign Types:

- Standard pictograms or symbols (23 choices available)
- Room number signs
- Copy signs
- Braille cell slugs
- Lift signs and symbols
- Custom signs
- Floor maps
- Signs with holders/framing units

All signage plans shall be submitted to the Owner for administrative review and approval, and a sign permit shall be obtained from the relevant authorities for all approved signage.

2.1.6 Entry

Access driveways shall be sufficient in number to provide safe and efficient movement of traffic to and from a site. Vehicular access shall be provided from side streets, adjacent alleys, and parallel streets whenever possible.

The building should feature multiple entrances. Multiple building entrances shall be positioned to minimize walking distances from cars, and facilitate pedestrian access from public sidewalks, and provide convenience where certain entrances offer access to individual retail outlets, or identified departments in a shop.

The building design shall provide design elements which give customers orientation on accessibility and which add aesthetically pleasing character to buildings by providing clearly defined, highly-visible customer entrances. Projected or recessed entryways, higher rooflines, changes in building material or color are some of the methods that shall be used to create this effect.

The number of entrances for the principal building shall be addressed at the preliminary development plan stage. Attractive well-marked pedestrian links between parking and buildings shall be provided. The connections shall be designed as safe, clearly marked and attractive pedestrian walkways across traffic lanes, landscaped areas and parking lots. Walkways should be shaded and landscaped. Where walkways cross traffic lanes, special design features should be used to increase safety for the pedestrian. Potential design features include: raised or textured pavement, curb extensions to narrow the travel lane or low-level lighting, such as a bollard light.

Removable bollards shall be provided in locations where emergency access may be necessary. Bollards shall be used to separate pedestrians from vehicular traffic areas and to light sidewalk surfaces. Bollard design shall coordinate with other streetscape furnishings.

2.1.7 Parking

Planning for parking facilities shall address the Functional/ Operational - as in providing for safe and efficient passage of the vehicle and driver. A well-planned circulation system that efficiently moves vehicles in a well-defined manner, while avoiding and reducing potential conflicts between pedestrians and vehicles shall be designed. Parking areas should provide safe, convenient, and efficient access for vehicles and pedestrians.

To minimize the impact of large areas of surface parking on the aesthetic character desired for a quality mixed- use development: All outdoor parking areas should be divided into smaller units to decrease visual impacts associated with large expanses of pavement and vehicles, and to facilitate safe and efficient pedestrian movement between parking and mixed-use development. Parking areas should be located on the sides or rear of the buildings with pedestrian connections between the parking areas. A landscaped separation should be used between parking areas and buildings to create a visual landscaped foreground for buildings.

Where parking structures are utilized, they shall be designed as an extension of the overall design concept incorporating similar materials and architectural detailing and should be designed to provide for first floor retail/ office use in commercial retail areas. They should be designed in a manner that would mask their purpose.

2.1.8 Landscaping

Landscaping in retail commercial development shall be intended, but not limited to making the environment physically more comfortable to the user, buffering or enhancing views, reducing noise, creating seasonal interest, assisting in water quality efforts and storm management, enhancing the public street appearance and enhancing the commercial retail development.

The development should have extensive landscaping of large parking areas,along streetscapes and for pedestrian- oriented open spaces which can be seen from the street and pedestrian-oriented areas. Landscaping should also help to define areas and separate areas, thereby bringing a human scale to these intense uses.

Unity of design should be achieved by repetition of certain plant varieties and other materials and by coordination with adjacent landscaping where appropriate. The choices, placement and scale of plants should relate to the architectural and site design of the project. Plantings should be used to screen, to accent focal points and entries, to contrast with or reinforce building design, to break up expanses of paving or wall, to define on-site circulation, to provide seasonal interest, and to provide shade. Landscape berms and/or continuous rows of shrubs to screen parking from adjacent developments or public streets are required.

Landscaping must be incorporated in the design of pedestrian areas along the building fronts. The use of raised planters, at least 30 cm in height for landscaping is strongly encouraged in retail centers where there are multiple tenants or large singular tenants. Parking areas shall have one tree per every four parking bays. There shall be sufficient trees, so that fifty percent (50%) of the parking lot is shaded within a five-year period. Parking lots shall provide landscaping next to buildings and along walkways, and landscape beds have a ninety percent ground coverage in five years.

Arbors or trellises supporting living landscape materials should be considered for ornamentation on exterior walls. Any such feature should cover an area of at least 95 m^2, and include sufficient plantings to achieve at least thirty percent (30%) coverage by plant materials within three years.

2.1.9 Site Utilities and Mechanical Equipment

All mechanical equipment such as compressors, air conditioners, antennas, pumps, heating and ventilating equipment, emergency generators, chillers, lift penthouses, water tanks, stand pipes, solar collectors, satellite dishes and communications equipment, and any other type of mechanical/electrical equipment for the building must be indicated on the architectural drawings.

Roof top mounted equipment shall be screened from the street and other buildings on all four sides by a structural feature that is an integral part of the building's architectural design. Mechanical equipment should not be located on the roof of a structure unless the equipment can be screened. The mechanical/electrical equipment should be clustered as much as possible. Roof top equipment shall be grouped and located so that it is not visible from the line of sight angle from the pedestrian right of way. Ground level mechanical equipment shall be screened with landscaping, berms and architectural walls using materials compatible with the building. Fencing materials are not allowed.

Figure 2.1.9 Roof Equipment Screening

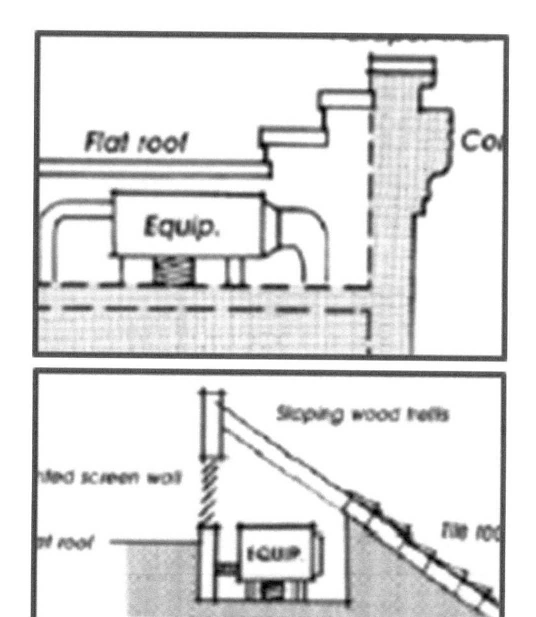

2.2.0 Loading Dock and Trash Storage Areas

Loading docks used for the arrival and departure point for large shipments brought to or taken from the building by trucks and vans. Loading docks are utilitarian spaces that should be designed for function and durability. However, it is also important that they are designed to ensure the safety and security of their users and the users of other nearby spaces. This space type must be able to accommodate large vehicles, forklifts, and pedestrian traffic.

Loading and unloading areas should be large enough for needed queuing but not allow parking. The loading dock should be designed in consultation with the Owner.

Loading docks serving the Conference and Exhibition Centre should be arranged so that the dock is continuous, providing access from each parking space to each hall division for maximum flexibility of use. One truck dock berth per 930 m^2 of exhibition space is the recommended ratio.

The dock surface shall be at the same floor elevation as the exhibition hall, to facilitate set-up/take-down of exhibits. In addition to the exhibition hall docks, dedicated food service docks, berths for trash dumpsters/compactors, a recycling station and a separate, enclosed food waste area are required. Trash enclosures shall be architecturally integrated into the design of the structure, at the rear of the building, and provide adequate space for recycling.

Dumpsters and trash shall not be visible above the height of the surrounding enclosure. Unroofed enclosure walls shall be a minimum of 2.5 m high. Trash enclosures visible over the 2.5 m walls from residential areas or public streets should be roofed. Typical features of loading dock space types shall include the list of applicable design objectives elements as outlined below:

Accessible

A ramp should be provided from the loading dock down to the truck parking area to facilitate deliveries from small trucks and vans. This ramp should have a maximum slope of 1:12 and comply with ADA/DDA Accessibility Guidelines, ensuring that it may be easily maneuverable for deliveries on carts and dollies.

Functional / Operational:

Loading docks should be located for easy access by service vehicles and should be separate from public entrances to the building and public spaces. Loading docks should be convenient to freight lifts, so that service is segregated from the main service lift lobbies. The service route from the lift should accommodate the transport of large items. Separate or dedicated docks should be considered for food service areas.

A single checkpoint shall be used by staff to receive all non-event and non-food materials for use in the facility. From here, deliveries should be routed internally to their destination or warehoused in assigned storage areas. Typical deliveries include dry goods and other consumables, office supplies and maintenance materials. All food service deliveries, including related paper products and other consumables, should be routed externally to a separate food service receiving area.

A receiving counter and vestibule should be provided with internal access to the main storage area. Allow clear access for forklifts from the loading dock into storage via roll-up door. Storage for general uses shall be assigned by the architect and should be distributed throughout the building.

Accommodating Vehicles:

Loading docks must accommodate vehicles used to deliver or pick up materials from the building. If the bed height of vans and trucks varies more than 45 cm, at least one loading berth should be equipped with a dock leveler. Typical docks are built 140 cm above grade level to accommodate most trucks.

Lighting:

Each truck position should be equipped with adjustable lighting fixtures for the illumination of the interior of trailers.

Edge Guards and Bumpers:

Loading docks should be protected with edge guards and dock bumpers.

Exterior Doors:

Easy access overhead coiling doors are required for loading docks. These doors should be able to close completely and lock after business hours. At least one well-lit personnel door should be provided in addition to the overhead doors.

Noise Exposure Mitigation:

Noise reductions in the dock and noise transmission out of the dock shall be important design considerations. Mass and limpness/flexibility are two desirable attributes for a sound transmission barrier. Unpainted heavy masonry walls provide mass. Absorptive acoustical surfacing will reduce the noise level in the dock but will have little effect on the transmission outside it. Noise levels in the dock should be moderated to promote communication among users.

Weather Protection for Users and Goods:

Open loading docks should be covered at least 125 cm beyond the edge of the platform over the loading berth to protect users and goods being unloaded. Based on the building location climatic conditions, dock seals should be used at each loading bay. Alternatively, consideration should be given to enclosing.

Staging Area:

A staging area inside the building should be provided adjacent to the loading dock. It must be protected from the weather.

Monitoring Ingress / Egress:

A dock manager's room or booth should be located so the manager can keep the entire dock area in view and control the entrance and exit from the building. The flow of circulation into the dock should pass this control point, and access should be restricted to authorized personnel. Security cameras should serve as a backup.

Emergency Egress:

Loading docks should not be used as emergency egress paths from the building. Additionally, staging areas and associated equipment must not interfere with emergency egress routes from the building.

Blast Protection:

Based on risk analysis for the building, the loading dock should be located so that vehicles will not be driven into or parked under the building. If this is not possible, the service area should be hardened for blast. Docks should be separated by at least 15 m in any direction from utility rooms, utility mains, and service entrances including electrical, telephone/data, fire detection/alarm systems, fire suppression water mains, cooling and heating mains, etc. Locate normal and emergency fuel storage areas away from loading docks.

Sustainable:

The entrances and exits at loading docks and service entrances should be provided with a means to reduce the infiltration of outside debris into the building. Maintaining a negative pressure in docks, relative to the rest of the building, will reduce infiltration and enhance indoor environmental quality. Carbon monoxide exposure from trucks and powered material-handling equipment, such as forklifts, presents a serious health risk to employees who work on loading docks. To eliminate or reduce exposure to the gas take the following steps: Consider using electric powered fork lifts or electric powered jacks; Install a carbon monoxide alarm system.

2.3.0 Life Safety

2.3.1 Overview

Fire protection for buildings and their occupants consists of various elements. This section is an overview of the International Building Code's (IBC) elements and integration of both passive and active fire protection to reinforce one another—and to cover for one another in case of failure of any one element.

For successful control, the active system, whether it is water, gas, aerosol, or foam, shall be capable of containing and acting on the fire while it is still a manageable size. The active and passive systems shall be capable of working together. The passive fire protection will provide the final opportunity to stop fire and smoke, and is crucial in confining fire to a manageable size for an active fire protection system to work on.

Where open atria consisting of a large open core extending through the center of the building from the lobby, having multiple levels of commercial occupancies that surround the atrium on the lower floors; with the remaining upper floors either used for residential apartments, offices, or other occupancies, facing the open core of the building through open corridors are designed, protection against fire and smoke exposure shall fully comply with all relevant IBC and local construction codes.

2.3.2 General Construction Requirements

The buildings should be of non-combustible construction as per Tables 503 and 601 of the IBC. The structural framing of the building should be determined by the range of systems permitted by the building code for the project. The structural frame and load bearing walls of the building should have 3-hour fire resistance rating. The floor and roof assemblies and shaft enclosures should have a 2-hour fire resistance rating.

The design, materials, construction and quality of roof assemblies, and rooftop structures should fully comply with the provisions of Chapter 15 of the IBC.

2.3.3 Fire/Smoke Separations

Internal fire/smoke spread through open stairwells or escalators, openings in otherwise enclosed stairwell or lift shafts, ductwork or "poke-throughs" (utility openings in floor slabs). such as vertical runs of electric trunk cable passing through many floors shall be prevented with fire- resistive structural elements, floors, ceilings/roofs, and walls to form an effective compartment, the construction assemblies (walls, floors, ceilings, etc.) should have the necessary fire resistance based on the amount of fuel load available in the compartment.

The enclosure and structural elements need to resist the exposure from a fire involving all the combustible material in the compartment. It is vital that compartments be properly constructed and maintained to preserve integrity throughout the life of the building. Furthermore, it is imperative that the compartment have continuity. Vertical subdivision must be continuous to exterior walls or other fire-resistance-rated walls. Horizontal subdivision must be continuous from floor slab to floor slab or floor/ ceiling fire-resistance-rated assembly. All openings need to be adequately protected and all penetrations sealed.

High rise structures require two-hour rated construction at Lift shafts, flues, linen/ waste chutes and pipe chases. Non- combustible construction is required at storage rooms and trash collections rooms. The parking garage should be separated from the residential area of the building by a 2-hour rated occupancy separation as per Table 707.3.10 of the IBC. It will also be a Type 1 Fire Resistive construction with floor and roof assemblies having 2-hour fire resistance rating. Such construction will allow an unlimited floor area for the parking garage as per Table 503.The occupancy separation between retail stores and the garage and between retail stores and residential units should be in accordance IBC required fire resistive construction.

External fire spread shall be prevented by the use of safe and acceptable EIFS (Exterior Insulation Finish System) designs that meet or exceed NFPA 285, Standard Fire Test Method for Evaluation of Fire Propagation Characteristics of Exterior Non-Load-Bearing Wall Assemblies Containing Combustible Components.

2.3.4 Fire Fighting Access

Fire apparatus access roads shall be installed and arranged in accordance with Sections 503.2.1 through 503.2.8, and Appendix D of the latest version of International Fire Code (IFC).

2.3.5 Means of Egress

Reference shall be made to Section 1.7.4 Means of Egress. The design team shall strictly follow the IBC standards to define the number of exit paths out of a space, and the number of exits out of the building.

Storage and storage closets are prohibited in emergency exit stairwells. If service areas open to staircases, they should be enclosed with one hour rated fire doors that fully comply with the latest version of NFPA 80, and fitted with smoke detector.

Section 403.5.4 of the IBC requires the exit enclosures to be pressurized in accordance with Section 909.20 and 1022.10 where the height of the building is greater than 23 m. Exit passageways serving as an exit component in a means of egress system should comply with section 1023 and 1028 of the IBC.

Rooms or spaces with an occupancy of 50 persons or more: exits should be provided in accordance with the current version of NFPA 101 Life Safety Code. When two or more exits or exit access doors are required, they shall be at least half (½) of the longest diagonal of the space that they serve. This requirement shall apply to each room, space or story that require two or more exits. If the building is provided with an automatic sprinkler system, the dimension between the exits can be reduced to one-third (1/3) of the longest diagonal. Stair enclosures that serve two or three stories shall be one-hour fire resistance rated. Enclosures that serve four or more stories must have a minimum two-hour fire resistance rating.

Fire alarm pull stations (manual call points) are required adjacent to all exit doors that lead directly to the exterior and at all stairwell doors, in major public areas/plant rooms and back-of-house areas.

Provide illuminated exit signs at each exit location and illuminated directional exit signage when the exit is not immediately visible. Emergency power is required at all exit signage in accordance with section 1022.9 and 1024 of the IBC.

Products, devices, and assemblies should be subject to laboratory testing criteria.

Table 2.3 – Occupant Load Design Factors (Based on IBC Table)	
Function Spaces	Floor Area in m² per Occupant
Space without fixed seats (chairs only – not fixed)	0.65 net
Standing Space	0.45 net
Unconcentrated (tables and chairs)	1.40 net
Business Area	9.30 gross
Commercial Kitchens	18.60 gross
Mercantile – Retail	
Basement and Grade Floor areas	2.80 gross
Areas other Floors	5.60 gross
Storage, Stock Areas	27.85 gross
Residential	18.60 gross

2.3.6 Fire-Rated Doors

Means of Egress doors shall meet IBC section 1008, and section 1017.2. Means of Egress doors shall be readily distinguishable from the adjacent construction and finishes so that the doors are easily recognizable. Doors in fire-resistive partitions should be self-closing with the following ratings as per Table 716.5 of the IBC, and fully comply with the latest version of NFPA 80. Doors in the means of egress in buildings with an occupancy in Group A, B, M, and R and doors to tenant spaces in Group A, B, M, and R shall be permitted to be electromagnetically locked if equipped with listed hardware that incorporates a built-in switch and meet the requirements below:

- The listed hardware that is affixed to the door has an obvious method of operation that is readily operated under all lighting conditions
- The listed hardware is capable of being operated with one hand operation of the listed hardware directly interrupts the power to the electromagnetic lock and unlocks the door immediately Loss of power to the listed hardware automatically unlocks the door
- Where panic or fire exit hardware is required by IBC section 1008 1 10, operation of the listed panic or fire exit hardware also releases the electromagnetic lock
- Fire door assemblies required to have a minimum fire protection rating of 20 minutes where located in corridor walls or smoke barrier walls
- Fire door assemblies in interior exit stairways and ramps and exit passageways shall have a maximum transmitted temperature rise of not more than 250°C (450°F) above ambient at the end of 30 minutes of standard fire test exposure

- Fire-rated doors and frames should be independently certified as fire rated in the State of Qatar by a national fire body authorized to certify such systems. A proprietary metal seal/sticker should be attached to the edge of the door and frame indicating the rating.
- Doors in smoke partitions shall be self- or automatic-closing by smoke detection in accordance with IBC Section 716.5.9.3

2.3.7 Ducts and Air Transfer Openings

Each opening through a fire wall shall be protected in accordance with IBC section 716.5, and shall not exceed $15m^2$. The aggregate width of openings at any floor level shall not exceed 25 percent of the length of the wall.

Fire dampers should be provided as per section 717 of the IBC. All penetrations in fire-resistive construction should be protected with through penetration systems approved by Underwriters Laboratories or equal.

2.3.8 Finishes

All materials used in all public areas should be inherently flame retardant or pre-treated to meet or exceed minimum requirements for a Flame-Resistant rating in accordance with latest edition of NFPA 701 (Methods of Fire Tests for Flame Resistant Textiles and Films), NFPA 255 (Test of Surface Burning Characteristics of Building Materials) and local fire and/or building codes whichever is higher.

Vinyl wall coverings should be CCC 408A type II at public areas. Type I wallcoverings is required for Residential Apartments. All wallcoverings should comply with local regulatory and fire authorities.Hard surface tile or stone should be non-slip finish with a minimum static coefficient of friction of 0.55 - 0.6 per ASTM C 1028.0R DIN 51097.

2.3.9 Miscellaneous

Gravity Waste Chutes installed to facilitate central trash collection should be equipped with lockable, self-closing and self-latching UL approved doors. Chutes should also be protected with automatic sprinkler protection.

Provision should be made for ignitable materials, paint, solvents, and fuels to be in containers and stored in approved UL/FM listed lockers or paint cabinets. Manual or automatic fuel/power cut off system should be provided in kitchen areas, and areas housing oil/gas fired boilers.

Special hazards should be reviewed by a qualified engineer. These hazards include, but are not limited to electrical transformer vaults, large gas, fuel oil or chemical storage facilities, and extraordinarily large IT and/or PBXrooms.

SECTION 3.0

Building Site

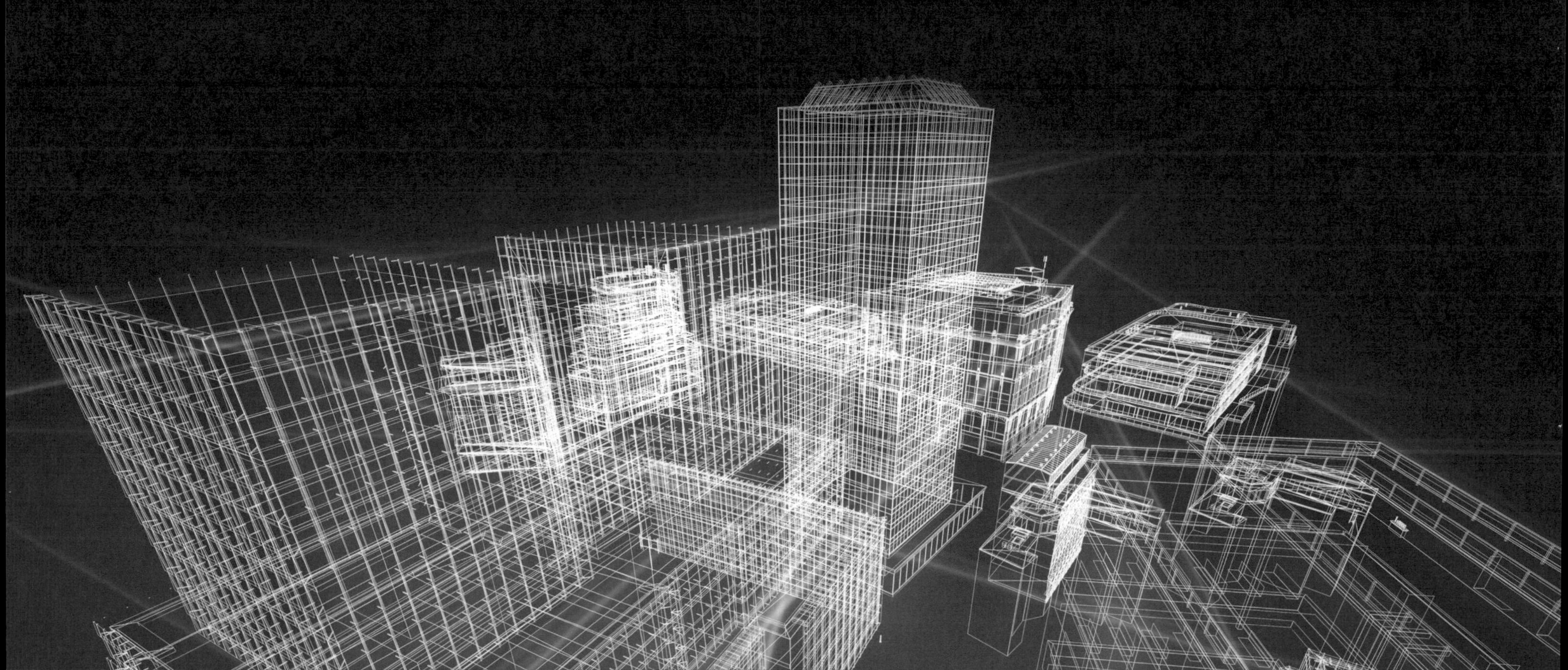

3.1.1 Site Survey

A detailed site analysis shall be undertaken to verify site information provided by the Owner, and understand the features of the site essential for the design stage. The site analysis shall record and evaluate information on the site and its surroundings, and to use this evaluation in the design response.

Site surveys shall include, but not limited to:

- Geological and geotechnical.
- Topographical surveys,
- Environmental Contamination surveys,
- Ecological survey,
- Traffic and transport.
- Local climate.
- Flood risk.
- Air quality.
- Acoustics.
- Boundary surveys.
- Electrical infrastructure and capacity.
- Gas network infrastructure and capacity.
- Foul sewers and drains infrastructure and capacity.
- Existing water supply infrastructure and capacity.
- Soil survey.

3.1.2 Orientation

The building's form, orientation and envelope construction shall be determined by considering appropriate passive design strategies. Passive design refers to the creation of building elements and configurations that take advantage of the physical environment of the site, such as climate data, building site's existing or potential topography with its landscape details, and potential optimization of natural on-site resources.

The project team shall use design simulation software for sun path diagram and shading model analysis to optimize the geometry of the building. During the process of establishing the orientation and form of the building, the following need to be addressed:

- Effective use of daylight
- Satisfactory building heat balance
- Utilization off seasonal solar gain and avoiding glare
- Minimum exposure to prevailing wind and wetting
- Minimum extent of noise penetration using appropriate areas as buffer zones
- The provision of a satisfactory visual environment with good personal control
- Efficient use of materials considering energy and without compromising adaptability or lifespan
- Appropriate use of thermal capacity

The orientation of the site shall provide important information to assist in siting of the building. This, when combined with the wind direction and sun path could establish the orientation of the building, and optimize the design. The orientation along with the sun path should determine the placement of rooms inside the buildings.

The design team shall protect natural resources using an ecology-based planning approach. A planning tool such as "Site Fingerprinting" shall be used in the design where protection of natural resources is the primary focus. This process will enable the design team to view, identify, and analyze the natural, built, economic, and social aspects of a prospective site.

Views onto and across streets and other public spaces shall be encouraged. For these frontages, the orientation of each building shall deal with issues of privacy. Views from one building into adjoining buildings are, generally not acceptable, and the design of new buildings is expected to limit intrusion into the privacy of existing properties. The location and design of buildings, and open spaces must be carefully orchestrated to maintain reasonable levels of privacy for adjacent development.

3.1.3 Wind Direction

To facilitate the design of a climatologically responsive building, the design team shall research wind direction, velocity and frequency.

3.1.4 Topography

The contour locations and spacing of contours will play an important role in the siting of the building. It is essential that to design the building along with the contours, integrating it into the design to prevent unnecessary cutting and filling of soil.

3.1.5 Building Placement

The architect shall locate the buildings parallel to the street and extend the base of the building the length of the site along the edges of streets, and open space. Provide greater building setbacks at strategic points or along the entire frontage, as appropriate, for architectural interest and to improve pedestrian amenity, including more space for tree planting, wider sidewalks, forecourts and publicly accessible open spaces.

Where applicable, maintain the character of existing soft landscaped streetscapes by providing generous setbacks for trees and planting.

3.1.6 Traffic Access General

The architect shall use high quality architectural and landscape design to emphasize primary entrances. It is essential to differentiate between residential and commercial entrances in mixed-use buildings. If a base building provides access to more than one tower or more than one use within a tower, ensure that the entrance to each is clearly identifiable, visible and universally accessible from the public sidewalk.

While it is generally desirable to have, a main entrance opening off the most important street bounding the site, this may not always be practical or permitted by local regulations off a main high-speed traffic artery. In such case, it should, subject to the Owner’s approval, be acceptable to feature multiple entrances facing less important public thoroughfares.

Vehicle access to the property should be as direct and obvious as possible. Devious, indirect and roundabout routes should be avoided, especially any routes which would cause the motorist to lose sight of the mixed-use- building, or which would require crossing heavy traffic lanes, without traffic signals.

It should be the responsibility of the Architect to consult with local traffic authority, not only to agree an acceptable traffic plan for access to the property, but to determine whether there are any plans for future traffic development in the area which might affect movements to and from the property.

Entrance Drive

If possible, the entrance drive from the public right-of- way should be under the complete control of the property management. Design a well-planned circulation system that efficiently moves vehicles in a well-defined manner while avoiding and reducing potential conflicts between pedestrians and vehicles.

Security should be incorporated into the most basic design considerations. The key site design concern should be access by pedestrians and vehicles. Security should provide for vehicles and pedestrians to be directed into specific patterns of approach through the site design. The points of access for pedestrians should be limited to provide a maximum amount of surveillance and control. Walkways should be set away from the building, and plant material and landscape features should not obstruct a clear field of vision around each building. Approaches to entrances should be open but provide controlled access by vehicles.

Separate entrance facilities for pedestrians and vehicles may be considered. The separate entrance isolates everyone entering the facility for a security check and could serve as a barrier to vehicles attempting to get to the entrance.

The internal site vehicular circulation system shall minimize conflicts between inbound and outbound traffic. Provide through-lanes (public or private) to minimize vehicle turnarounds. Separate service entrances, served from secondary rear streets shall be provided with direct access to the service areas located away from shopping areas and planned amenities.

Construction:

All carriageways, within the property curtilage should be properly graded, drained and constructed in accordance with applicable local regulations and specifications approved by the Owner. No carriageway should have a gradient greater than 10%.

In cases where the length of the carriageway may incline drivers to excessive speed,"sleeping policemen," (humps in the carriageway) should be constructed to enforce a reasonable limit of speed. These should be painted in a contrasting color to be readily discernible.

Hose outlets should be provided in convenient locations for the hosing down of carriageways.

Wayfinding Signage:

The design team shall study the circulation patterns for vehicular and pedestrian traffic, to understand each visitor sequence. Based on the site and circulation analysis, a holistic wayfinding approach shall be developed for the entire property that will address each customer type and sequence as part of one comprehensive and flexible system.

The following general guidelines shall be followed for the signage program:

- Design signs for existing climate conditions, and to meet local codes.
- Specify sign illumination fixtures as required for even light distribution and visibility of sign faces.
- Standardize sign design and messaging for all signage in each area.
- Limit information on signs to the essentials, based on the location and viewing audience.
- Standardize terminology and use of symbols on all signage in each area.
- Use English and Arabic only as the message standard for signage; continue and expand use of international symbols.
- Establish information hierarchies and distinct viewing zones for wayfinding and other forms of visual communication.

I. **Taxi Stand**

A taxi stand of adequate capacity should be conveniently located with respect to the public entrances, with appropriate signaling arrangements for calling taxis to those entrances. Where the taxi stand is located on the property, provision should be made for a shelter for waiting drivers, including a covered seating area with a bench or benches and minimum toilet facilities (WC urinal and lavatory), not accessible to the public.

II. **Valet Desk**

A valet desk should be provided adjacent to the primary entrance vestibule, convenient for visitor access and shielded from the elements. Provide a glass transaction window when the valet desk is located in a separate room, to allow visual connection between valet attendant and customer.

The face of the valet desk should be finished with natural wood, tile, stone or approved decorative material.

3.1.7 Parking & Garages

Parking should be located well away from the building structures, and controlled parking may be provided for selected persons. Strict setbacks from the building structures should be observed for all vehicles.

Parking areas should provide safe, convenient, and efficient access. They should be distributed around large buildings to shorten the distance to public sidewalks and to reduce the overall scale of the paved surface. If the building is located closer to streets, the scale of the complex shall be reduced. Provision should be made for the parking of customer vehicles in the number called for by the Fact Sheet.

The Architect should be responsible for ensuring that this number complies with requirements of local regulations. A reasonable number of these parking spaces should be provided near the main entrance of the mall, for the convenience of persons requiring space for temporary standing when awaiting customer pickup. The remaining spaces designated in the Fact Sheet will be for customers arriving by car, requiring parking facilities for a longer period. Whether such parking facilities are to be in the open, covered, or completely enclosed, will depend on the space available or on the climate and will be set forth in the Fact Sheet.

In areas subject to very hot solar radiation, adequate shading for cars should always be provided. In all parking areas, driving lanes and individual parking spaces should be clearly delineated by appropriate painted lines in highway quality paint Primary circulation paths should avoid excessive steps or level changes to reduce potential tripping hazards and facilitate circulation for all potential users, including strollers and wheelchairs. As required by the IBC, accessible parking is required wherever surface or structured parking serve accessible buildings. The current ADA/DDA requirements shall be applied.

Table 3.1 Recommended Parking Ratios for Preliminary Design

Type of Use Served by Parking	Spaces
Shopping Malls	**4 – 6 per 93m² gross area**
Retail	**2 – 4 per 93m² gross area**
Restaurants	**10 -25 per 93m² gross area**
Residential Apartments	**0.5 - 2 per 93m² gross area**
Office Building	**0.5 - 3 per 93m² gross area**
Convention Centre	**20 per 93m² gross area**

Covered or enclosed parking areas or garages should be provided with control facilities consisting of an automatic parking ticket dispenser and entry gate, together with a collection booth and automatic exit gate. Provide Time Stamp, Office Telephone and air-conditioning in the collection booth.

If the garage is to be operated by a concessionaire, a separate toilet and office facility will be needed.

Covered or enclosed parking areas or garages should be provided with signage that clearly indicates connectivity between the garages and the mall. Level/Row Column Marker Signage shall guide the customers find their vehicles upon their return.

3.1.8 Landscape

General

All grounds should be landscaped and planted with well- developed trees, shrubs and flowers indigenous to the area, or which can be expected with assurance to live in the climate with normal care and attention. All plants should carry a warranty for a period of one year, with at least an initial maintenance contract arranged as part of the general contract for the project with the vending nursery to insure proper care and instruction to staff landscaping personnel during that period.

The design of the grounds and the planting should be performed by a qualified landscape architect approved by the Owner, working under the direction of the Architect. The planning of the grounds should be accomplished and the actual work should be performed on a time schedule which will insure that, at the time of the mixed-use building complex opening, the entire site will present an acceptable appearance of completion to the first visitors.

The architect should establish in the construction contracts the requirement that all contractors and sub- contractors cooperate fully to this end. The designer should preserve the natural features of the landscape, and design additional features to work in harmony with the environment.

3.1.9 Security and Control

Depending on the location of the project and site conditions, it may be necessary to enclose the entire grounds of the building complex, or provide fencing on one or more sides of the site. Unless expressly specified by the design brief, it should be the responsibility of the architect to investigate the need for such security precautions and make recommendations to the Owner.

Fencing arrangements, where required, should be adequate in extent, height, strength and permanency. Fencing should be decorative iron fencing, solid walls with decorative iron inserts, or other approved materials.

All materials and architectural details should coordinate with other building architectural features. Planting and fencing should be used also to screen and protect unsightly installations such as outdoor transformer stations, liquid gas tanks, etc.

In planning the grounds and landscaping, provision should be made to reach all parts of the grounds by truck,for maintenance purposes.

3.2.0 Plumbing systems

An automatic irrigation System should be employed ensuring continuous growth. The irrigation system should apply a consistent, even, measurable amount of water to the landscape over a period of time. It is necessary that the system design consider water cost and conservation, long term durability and maintenance cost, safety issues, aesthetic issues, and site-specific requirements. The irrigation system should be designed to adequately water the landscaped area in the "worst case" condition. This is usually midsummer when the average daily temperature is at or near its highest for the growing season or when humidity is averaging its lowest percentages.

Bubblers and drip irrigation devices should be included in the system where low volume irrigation is required, and should be separately zoned, and valved independently from rotors or spray heads.

Adequate drainage throughout the grounds should be provided, with particular care to avoid any standing water on terraces or paths designed for guest or service use. See Plumbing Section 23.1.4

3.2.1 Site Amenities

General

Similar to site design and building architecture, site amenities such as courtyards, site furniture, and hard landscaping features contribute to the overall tone, image, and style of the mixed-use project.

Outdoor spaces shall play a significant role in the development of the site plan and shall be designed as "outdoor rooms" that can be used for play, recreation, social or cultural activities. The architect shall avoid undifferentiated, empty spaces. Outdoor spaces shall be appropriately scaled for the intended use and be designed to include safety and security measures.

Hard landscaping materials should include all areas of paving, gravel works, fencing, water features, outdoor furniture etc. Walkways and decks should provide convenient circuits between parking areas and building entrances.

Decorative items such as lamps, benches, planters accented with stone, brick, together with focal elements such as sculptures, art, or water features shall be incorporated into courtyard and plaza design.

The landscaped courtyards and outdoor spaces should be created and shaded with trellises and tree canopies. This filtering of daylight and sunlight will allow for a connection with nature without the need for enclosed air- conditioned spaces.

Both private and semi-private outdoor spaces shall be incorporated in mixed-use developments. Private outdoor courtyard areas for residents only are strongly encouraged. Semi-private plaza areas for visitors shall also be provided in areas adjacent to the retail/commercial uses.

The design of trash receptacles shall coordinate with other streetscape furnishings. In residential or mixed-use developments provide access to secure outdoor play space and equipment for family sized units. Finishes and Weather Protection. Graffiti resistant material and/or coating shall be required to retain the furniture's attractiveness.

Outdoor terraces for visitor use should have well-finished, even paving, not conducive to tripping, ankle-turning, or unsteady tables and chairs. Paving should not be slippery when wet. All paving should be sloped for natural drainage or to surface water drains. In areas where breezes or winds of uncomfortable temperatures or intensity can be anticipated regularly, at frequent intervals, or at certain seasons of the year, exposed terraces should be provided with permanent or knock-down" screens which will permit the use of the terraces at times when the weather is otherwise agreeable.

Where the building program calls for a "covered" terrace, a permanent structural cover should be provided, unless a removable cover is specifically called for.

SECTION 4.0

Facility Description and Functional Relationships

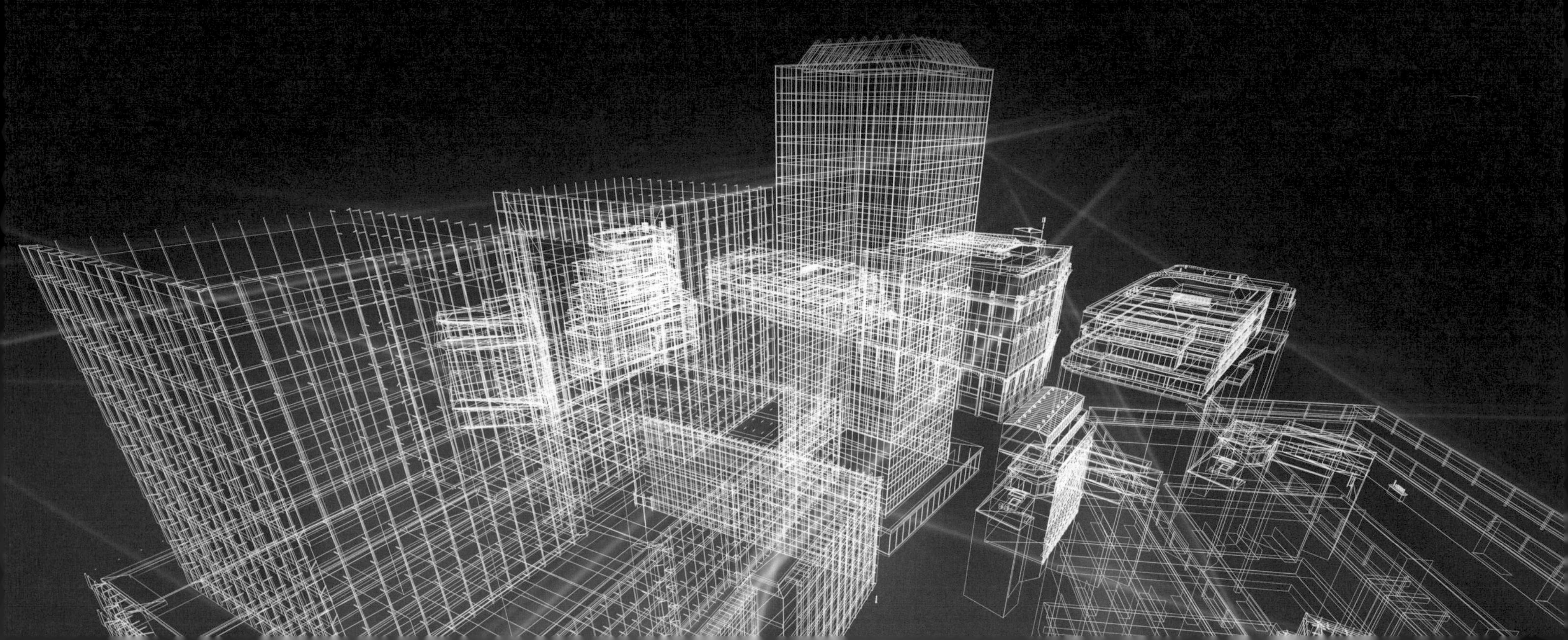

4.1.1 Space Standards

General

The hallmark of a good space plan/interior design is to create a space that is both functional and aesthetically pleasing. The space standards which are given in the following paragraphs of this section are guidelines based on established norms for a Mixed-Use building with full support facilities. They represent a fairly well balanced operational and economic relationship between the individual functions and external facilities which make up the building complex of an up-to-date Mixed-Use building complex of its class.

The standards serve, therefore, as the essential point for establishing the basic design criteria for a prospective new building; or as a yardstick for measuring the adequacy or otherwise of established buildings, when contemplating major design changes as part of a rehabilitation program. However, they should be subject to critical evaluation in the light of the differing circumstances which inevitably confront each and every mixed-use building complex. Each will have its own special needs, dictated by its economic environment, site planning constraints, local traditions, labor practices, support facilities and market demands.

Variations in environmental and economic factors should affect the planning criteria. They are only a few of the various influences which, revealed by initial planning and feasibility studies, may warrant variation of one or more specific standards. These standards will be used during the preliminary design program for the preparation of schematic plans only. All dimensions and areas used are metric (mm/m/m²). All areas relate to net useable space, i.e., the space between enclosing walls or partitions. They do not include linking corridors, structural columns, wall thicknesses, ducts, etc. As a "rule of thumb", an allowance of 25% may be added to the total net areas for grossing up to total building area for preliminary cost estimation purposes. This also acts as a control of economic planning development. Full details on operational, architectural and interior design criteria on which the space standards are based will be found in the relevant references.

4.1.2 Accessibility

The design team shall fully comply with IBC scoping requirements for accessibility. The latest version of Accredited Standards Committee A117 on Architectural Features and Site Design of Public Buildings and Residential Structures standards shall be applied to all buildings and facilities.

An accessible path to the building and through the building shall be accomplished by providing an "accessible route". An accessible route shall be a continuous unobstructed path that complies with the IBC and A117.1. It can be a walkway, doors and doorways, ramps, curb ramps or lifts. These components of the accessible route could be used by a person in a wheel chair or by a person with other form of physical disability. Persons with physical disabilities shall be able to move through the building and enjoy the services the same as anyone else.

To provide equal accessibility in assembly areas, accessible seating areas shall be provided. Total area allotted to seating in dining areas shall be accessible. All floors that are accessible shall be provided with an "accessible means of egress". The general intent is thatthe building shall be constructed in such a way that persons with physical disabilities are protected until the Civil Defense Department can reach them and rescue them.

4.1.3 Environmental and Architectural Parameters

Basic architectural constraints and parameters of an individual project shall be established during the initial programming phase, so that the relationships between the client's needs and qualities of physical space are known from the outset.

Structural systems, construction materials, fenestration types, building shape and configuration, and architectural design and detail have a major effect on space planning decisions. Architects and structural engineers shall be in close consultation to determine the degree of flexibility and openness that can be created; The length of a building's structural spans (bay size) would be of essence, as small bays will restrict partition and furniture placement and limit traffic flow.

A combination of structural system and materials factors will govern, where, and how door and window openings can be made in both interior and exterior walls. The relative simplicity or complexity of a building's shape or configuration will determine its usefulness for a given set of planning requirements; complex exterior wall configurations and peculiarly placed stairwells and lift shafts can render a building unsuitable for a particular use.

In space planning terms, the design program shall be a written document that qualify and quantify the client's or the user's needs for a given project. Additionally, the design program shall be accompanied by relationship diagrams that express physical planning relationships accurately.

4.1.4 Criteria Matrix

The design process will be determined by the Owner's precise needs. A Design Brief 'Brief to Architect' drawn up by the Owner appointed specialist consultant, should clearly list down and describe in a simple & easy to understand manner, all the requirements & expectations from the design.

The architect shall develop a Criteria Matrix (CM) that will visually organize information of a variety of factors of the design program in a condensed form. For example, the CM will include daylighting and shading requirements for some of the spaces in the column titled "Sustainable Factors." A sample format is shown in Appendix "II".

The space planning analysis shall focus on the concepts of zoning and isolation. When user and spatial needs are being organized, as in the development of the CM, acoustic needs for privacy, isolation, and absorption shall be identified. The design team should create quiet and noisy zones ((often coinciding with private and public zones) during the process of defining the best possible inter-relationship between the various functional areas (bubble diagramming or block planning phase).

Sustainability considerations shall be a part of the bubble diagramming process. The following shall be thoroughly investigated and applied:

- Maximizing the use of high-quality diffused daylight
- Minimize the heat that comes with direct sunlight by adding shading on the outside of the building

4.1.5 Planning Brief - Commercial Areas

Every aspect of Mall Development is critical, ergonomic, scientific, requirement driven & process driven there by creating a niche for every segment of Mall development to be of great importance. The Design Planning phase in the life cycle of a Mall requires to be professionally managed by a team, which is technically well versed with all facets of Design, Leasing & Operations.

Based on this philosophy, a dedicated team of Retail Design Architects and Retail experts with in-depth understanding & experience Mall Design, understanding of the technicalities of site Circulation, Zoning & Planning must be appointed to advice the Owner.

The number of entrances for the principal building shall be addressed at the preliminary development plan stage. It is essential to have a special team or an expert Mall Advisor to conceptualize, efficiently manage & coordinate the design planning phase and have an organized approach towards Mall Design & Planning. A "Merchandising Plan" shall be devised in which primary store units, or 'magnet stores', would be so disposed that visitors to them would be led past secondary units, thus maximizing trading opportunities, turnover and rental levels for the Mall as a whole. The logic of merchandising planning, with its terminology of 'magnets', and its quantification of the successful plan in terms of rental yield, have led to a simplification of planning patterns into several generic types based upon the number of major magnets to be incorporated.

Thus, a center with a single magnet, would adopt a centripetal form, with incoming shoppers from the surrounding parking areas drawn down approach malls to the central store. Since only a proportion of the incoming traffic would pass down any one approach mall,these would be kept short, with the secondary traders concentrated as far as possible around the central core area where all shoppers might be expected to circulate. A much greater area of prime frontage for secondary traders could be created with two magnets, located at each end of a 'dumb- bell' plan with a central mall down which shopper would be drawn. In the same way, three department stores could induce a linear or a 'T' or 'L' shaped primary mall route. The design brief should provide information about the desired circulation pattern, in a manner that the customer is effortlessly guided through the Mall, as if in auto mode. Right from entry until the journey is complete utmost care should be taken to ensure that customers journey is enjoyable.

A good design should enable a shopper to find their way in, through and out of a Mall. Points of conflict must be minimized. It is generally accepted that the maximum distance a shopper will walk is about 200m. The Mall width shall be such that it will not discourage the shopper from crossing over to shop on the opposite side. The architect must be fully explained the proposed leasing concept. These details are generally provided by the Mall Advisor or Leasing experts. This brief should provide a clear understanding of the basic zoning and its related requirements. A list of the required locations of the different format shall be defined e.g. fashionableclothing boutiques, a specialty shop which deals mostly in sports and leisure goods is required on a specific floor, a wedding dress boutique on the first floor, Anchor stores on ground floor etc. This data is necessary to give clarity to the architect of the different services & infrastructure requirement of each of the format so that they can plan accordingly.

4.1.6 Retail Shop Design

Where multiple-tenant spaces are incorporated into a building, individual tenant spaces shall be located within the building bays. This shall be achieved by any of the following:

- Placing a column, pier or pilaster between façade elements.
- Applying vertical slot or recess between façade elements.
- Providing variation in plane along the building wall.
- Varying the building wall by recessing the storefront entrance or creating a niche for landscaping or pedestrian area.

The design of the entrance to a store should entice the customer in and give a view of the products beyond the threshold. It should be a new element fitted as part of an overall scheme. The customer should also be able to see past the display windows, allowing transparency and interaction. The tenant’s storefront design shall place emphasis on proportion, scale, color and detailing to enhance the quality and image of the Shopping Complex.

Tenants and their Architects/Store Planners are encouraged to design storefronts that explore creative uses of merchandising, lighting and signage. The interior of each demised premise should be consistent with the design concept or image created by the storefront. Three dimensional storefront designs with large amounts of glass emphasizing show windows and display cases is highly encouraged. These design goals can be accomplished through close attention to detail, use of high-quality materials, good craftsmanship and innovative design.

A Design Control Zone shall be established in all Tenant premises. The Control Zone shall extend a minimum of 1.25m into the Tenant leased premises, measured from the store closure line. The Owner shall have absolute right of approval overall design signage, and materials within this zone.

The Design Control Zone shall include display windows, retail graphics, display fixtures, signs, materials, finishes, color and lighting in front of the Design Control Line. The retail designer shall plan out the circulation within the retail space, taking into consideration the design guidelines and principles of the scheme alongside the structural nature of the interior. Space for support areas such as fitting rooms, staff/customer consultation areas, and areas that house the functional elements aside from selling, shall be shown in the space layout scheme. The functional elements shall consist of areas that support the running and managing of the store on a daily basis and provides essential areas for storage and staff facilities. The fitting rooms shall be separate spacious cubicles with mirrors on all sides, a fixed seat, with hooks to hang clothes, and a solid lockable door.The figures shown in Tablen4.1.6 are based on usable square meterage, which in the language of leasing includes the area within the boundaries of the leased space.

Table 4.1.6
Retail Store Space Planning Thumb Rules

Building Use/Type	Per Unit	Area Basis
Large Stores	2.8 – 4.65m^2 per person	net
Store Public Circulation/ Activity zone	91.4cm -130cm minimum	net

4.1.7 Restaurant/Food Tenants

The topic of dining and food service covers a vast array of facility types and operations from fast food kiosks to fine-dining restaurants. Since there are multiple design approaches to achieving the basic facility function, certain basic operational decisions must be made prior to initiating design:

- Number of people to be served
- Meal schedule and duration
- Food delivery and eating methodologies
- Any additional functions accommodated in the specific facility

The food delivery and eating methodologies affect the size and layout of the facility. Dining facilities may accommodate more than one of the following methodologies:

- Serving Line or Station.
- Customers choose from pre-determined options off of serving lines or stations such as hot food line, salad bar, deli bar, pizza bar, taco bar, etc.
- Food may be packaged for consumption in the facility or for takeout.
- Customers order items for custom preparation.
- Food may be packaged for consumption in the facility or for takeout. Takeout. Customers choose from assorted prepackaged items that may range from sandwiches and pizza to full meals. Payment is typically a la carte.

Once the operational considerations have been determined along with other standard facility planning factors, the following spaces shall be scoped and laid-out:

- **Entrance Lobby**. The size is primarily determined by the number of customers to be served.
- Queue. The queue is the space between the entrance lobby and the serving area and is determined by the serving capacity, the serving methodology.
- **Serving Area**. The serving area accommodates ordering and delivery of food to patrons and is determined by the food delivery methodology. The design of the serving area impacts the serving capacity and must be coordinated with the queue and dining area.
- **Cashier Station**. The cashier station accommodates patron payment, and the configuration, location and number of stations are determined by the number of people served, food delivery methodology, and the payment style. Payment options (cash, credit, pre-paid meal cards) must be determined prior to design.
- **Dining Area**. The dining area accommodates customer eating and relaxation. It is determined by the number of customers to be served and meal schedule and duration as expressed by turnover/serving capacity and seating capacity. The design must also be coordinated with the food delivery methodology and bussing approach.
- **Turnover/Serving Capacity**. Turnover is the number of times a dining area seat is occupied during a given meal period. Turnover drives the serving capacity, which is the number of customers served within the set meal duration. The serving capacity shall be used to size

the functional elements of the dining facility to ensure that the required number of customers can be served in the meal duration.

- **Seating Capacity**. Seating capacity is determined by considering the required serving capacity and the serving methodology. The seating capacity shall be used to size the dining area.
- **Kitchen and Preparation Areas**. The kitchen and all food preparation areas shall be determined by the number of people to be served, the food delivery methodology, the menu, the bussing style and the storage capacities.
- **Dish/Pot-Washing**. The dish and pot-washing areas shall be determined by the number of customers to be served, bussing considerations, the food delivery methodology, and the menu.
- **Storage**. Storage areas accommodate stocks of subsistence (consumables) and non-subsistence, e.g., tableware, cleaning supplies. The areas shall be determined by analysis of the menu, the number of customers to be served, and the defined delivery cycles.

Food Tenants within the Shopping Complex should offer a festive and inviting atmosphere for mall patrons. A unique, high-level design is expected from each Food Tenant to create a warm and inviting environment for customers. Food and Restaurant Tenants are highly encouraged to define their individuality by utilizing innovative space layouts.

A Design Control Zone shall be established in all Tenant premises. The Control Zone shall extend a minimum of 1.25m into the Tenant leased premises, measured from the store closure line. The Owner shall have absolute right of approval overall design signage, and materials within this zone. The Design Control Zone shall include display windows, retail graphics, display fixtures, signs, materials, finishes, color and lighting in front of the Design Control Line.

All tenancy leased areas must include the Front of House & Back of House, Dish-Wash areas, restricted areas & its circulation like service corridors. All food service counters must be located beyond the Design Control Zone and must meet the counter requirements for over- the-counter Food Tenants.

Figure 4.1.6 Mall Functional relationship

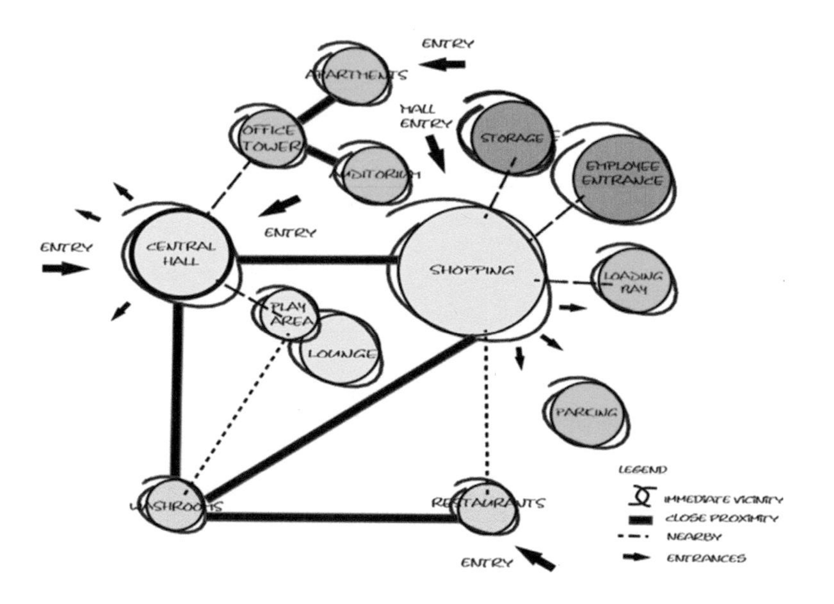

Seating for restaurant patrons must be incorporated within the lease space and restaurant tenants must dedicate a minimum of 40% of their leased space to seating. Where a Food Court is planned, the tenant must dedicate a minimum of 60% of leased space for seating. Outdoor terraces shall be considered as part of the Food Court layout.

Kiosks shall be located at the main entry points into the food court equipped with full service & wet points). All the Food Tenancies should have a service corridor, at least 1.5 m wide running behind these shops.

Storage areas / Changing Rooms / Locker rooms for the Restaurant and Food Tenant staff shall be planned suitable during planning stages. Also, the designer shall plan support areas to accommodate staff and administrative needs, and any other facility functions. Trash and garbage removal and recycling can have a significant impact on Dining Facility design and systems should be determined prior to design.

The design team shall organize each facility to foster efficient flow of people, materials, and work activities. It is essential to visually and acoustically separate customer functions from food preparation and cleaning functions. The relationship among the various storage, preparation, cooking, serving, and cleaning functions must be carefully studied to optimize work flow and efficiency. Keep travel distances short and minimize crossover of circulation paths. Plan for various serving styles.

The dining area represents the conclusion of the customer process of arrival, queuing, serving, and payment. To the extent possible, separate seated patrons from the congestion and movement of arriving and departing customers.

Type	Per Unit	Area Basis
Restaurant	$1.4m^2$ – $1.7m^2$ per seat	net
Food Court	$1.5m^2$ – $1.7m^2$ per seat	net
Counter Service	1.7m – 1.85m per seat	net

4.1.8 Conference/Meeting Facilities

Because of the large size of the Conference/Meeting Facilities and its urban context, these guidelines anticipate the facility having multiple entrances. While one major entry may be part of the architectural signature, primary entrances should provide access to registration and public circulation that welcomes visitors. These transitional areas should immediately orient attendees to the rest of the building with an impression of openness and indirect natural light. Glazed walls between the arrival zone and entrance concourses will reinforce the attendees' association with an event.

Architectural features should balance the grandness of the lobby with human scale details. These spaces should be configured to facilitate security control points into the auditorium, exhibition halls and meeting room blocks which can be adapted for both single and multiple concurrent shows.

Auditorium spaces shall be designed to accommodate large audiences. As such, they tend to have wide spans and are multiple-stories high to accommodate seating, sightline, and acoustical requirements. Multi-purpose halls that support everything from large meetings, presentations, and performances theater to music are the most common auditoriums. These spaces typically follow a proscenium theater design but can change as the requirements of the performance changes. Typical features shall include a fly loft and rigging system for lights, acoustical panels, curtains and stage scenery. A thrust stage and orchestra pit are common features.

The multi-purpose auditorium shall be designed to have a seating capacity of 1500 – 2000 seats, and have the following characteristics:

- Acoustical shells used to enhance acoustics in stage house and audience chamber.
- Acoustical clouds and panels used to support acoustics in audience chamber.
- Ceiling height that provides sufficient cubic volume for desired reverberance.
- Side walls shaped to reflect sound toward side of audience.
- Incorporate ways to vary strength and reverberance onstage and in audience chamber (such as adjustable absorption and coupled volumes).

Quiet so intruding noise does not interfere with performances. Enclosing constructions that isolate unwanted sound of adjacent spaces and outdoors. HVAC systems serving auditorium shall not produce disruptive noise.

The amount of space required for an auditorium depends on several factors, however the figures below provide an approximate guide. The calculations shall be based on a specialist consultants design brief. This type of space should be designed for performances and seminar-style gatherings and shall include a fixed stage and fixed seating.

The floor should be sloped or tiered to provide better sight lines. Provide seating that includes writing surfaces to facilitate lecture note-taking.

(a) **Pre-function Areas**

The gathering space outside of the auditorium and meeting rooms shall be primarily used for pre- and post- event social networking by attendees. Pre-function spaces outside of the meeting rooms should allow for tabletop registration at individual room events. Refreshments such as water, soft drinks, coffee and snacks are typically served, requiring access points to service corridors.

The design team shall allow one-quarter (¼) to one-third (⅓) of the adjacent room size to be used for pre-function and circulation, assuming all rooms are not filled. Alcoves for washrooms should be directly accessible from pre- function spaces, and located convenient to auditorium and meeting areas and vertical circulation cores.

(b) **Exhibition Halls**

The exhibition area is the largest single component hosting event activities of the Conference and Meeting facility. The space shall be configured to provide flexibility to handle simultaneous multiple events typically in different stages of the event cycle (move-in, event and/or move-out) while minimizing conflicts between events. All exhibition space shall be on one level. The configuration must permit direct access of exhibit materials to the floor through dock berths, and by means of ramps to allow over-the-road trucks to drive onto the exhibition floor. Exhibition design layout is the distribution of the elements of a program on a selected site to achieve sound functional relations with different functions, such as entrances, exits, landscapes, water bodies, buildings, transportation and wait stations.

There are general site conditions that are critical to the success of Exhibition Halls. The areas of the site should be divided in line with the type of the assigned service to each zone. Layout of the exhibition hall shall provide separate paths for attendees and service personnel. The intent is to minimize conflicts between the different traffic flows through the building. Paths may be separated vertically, horizontally or in a combination of both. The aim of the perfect design is to unite the movement of people in a way to enable them to see the exhibition easily without misleading the road or feel bored or tired. The specialist consultant should consider changes that may occur in the expected movement, to prevent the resulting gathering caused from people slowing and their curiosity.

There shall be two walk lines: limited (specified) line and unlimited line. The limited walk line: used if the goal of the exhibition is to provide a sequential topic and everyone should see everything. It should consider the following:

The limited space should not exceed 100 Meter, providing free places to avoid feeling of unexpected implementation with the diversity of the surrounded axis. The unlimited walk line: It followed in most of exhibitions, which do not need this sequence, such as commercial markets. This in the free projection to give the opportunity for the visitor to walk around. This kind takes many forms: It can be in a form of a sequence of showrooms bound with lanes. The routes should not be similar in front of the viewer so did not feel he strayed the way or that he did not see everything. Also, to avoid the straight lanes in projection. The winding lanes are the best, offering excitement and change.

(c) **Grand Ballroom**

The grand ballroom space shall accommodate formal dining and assembly events as well as provide flexibility, allowing simultaneous activities. This space should be highly visible and have a significant arrival and entry procession. The ballroom and its pre-function will have a higher level of finishes than other meeting rooms, typically the highest finishes in the facility.

The Grand Ballroom should be divisible into three (3) equal sections. The areas given are based on normal needs for developed areas. Local market studies may show need for adjustments to the allocations in the tables.

This room should accommodate up to 750-1000 people comfortably at a single banquet, and features a finished ceiling with a minimum clear height of 10 m. The ballroom stage shall have access to loading docks or freight lifts for move-in of products, presentation materials, and set decorations.

(d) **Meeting Rooms**

Meeting room spaces shall be designed to accommodate a wide spectrum of event types and group sizes. These spaces shall be associated with each exhibition hall division, and arranged to accommodate simultaneous multiple events. Meeting rooms can be on separate levels, but should be related to exhibition halls and assignable to each hall division. For planning purposes, the size of each module shall be established by the specialist consultant based on a market study. Flexibility in sizing of permanent meeting rooms is required so that a variety is available to users.

Minimum ceiling heights should vary with the overall room size to maintain the right proportions for multimedia presentations. The specialist consultant is encouraged to consider alternate mixes and configurations of meeting rooms, particularly when creating a stronger link to specific exhibition hall divisions.

(e) **Board Rooms**

This suite serves as the before-and-after-event lounge for featured presenters and very important guests. It may also be utilized for special meetings or receptions for small important groups. High quality food and beverage service shall be provided from a small servery/bar area with service access to the main kitchen.

Provide a secure back-of-house entry to a securable passenger loading area with access for limousines. The suite should include three specific rooms, a reception area/conference room, servery/bar and a private washroom.

Flexible lighting schemes should adapt the space from board meetings to multimedia presentations and video conferencing.

(f) **Event Managers Offices**

During the individual event, as well as during move-in and move-out, these offices will serve as the administrative center for the event promoter. During an event, these spaces may be used as small meeting rooms.

An ideal arrangement for the offices shall allow direct access to the pre-function concourse, with controlled access both visually and physically to prime exhibition space. The most suitable location for these offices is on a mezzanine level near the exhibition hall entrance. Dedicated offices for each exhibition hall division shall be required. A flexible suite that spans hall divisions is also usable as a single office for larger events.

Office access should be from the concourse or registration lobby and should not require going through a security checkpoint.

(g) **Business Centre**

This space will be used to provide a variety of support services to exhibitors and attendees alike. Opening onto the public concourse, there should be a semi-private lobby with a service counter to handle customer requests. Space should allow for several high-tech working cubicles to be installed along one wall for accessing computer- based services that include word processing, scanners and printers.

(h) **Public Washrooms**

Public washrooms should be well distributed throughout the public circulation areas to serve the meeting and banquet spaces. At least one set of washrooms should be accessed from within the exhibition hall. Washrooms should be located to serve blocks of meeting rooms and convenient to the general circulation pattern.

(i) **First Aid Room**

A First Aid Room is required to provide a facility for the treatment of minor injuries to center employees, service contractor personnel, and event attendees. This space will act as a holding area while more seriously ill or injured persons are awaiting transportation to a local hospital or other treatment facility.

This space should be located between the public concourse and exhibition hall for use both during move- in/out and the events. Access to back-of-house corridor is required for removing ill or injured patrons outside of public view.

4.1.9 Back of House Support Areas

Loading Dock: Refer to section 2.1.9

Compactor/Recycling Station

Solid waste handling shall be planned to recycle corrugated paper, aluminum, plastics and glass. A corrugated paper compactor and baler will be required, as well as a truck bay or other convenient storage location in the loading dock area for recyclables. Refer to Section 2.1.9

Security Control Room

The Security Office and control room shall be located at the employee entrance to the facility. It will be the control center for the building security system and where an emergency response team will coordinate their activities. The office space shall be fully equipped with custom console for monitoring fire, life safety, and security systems.

The room shall be designed to sound-absorbing acoustic design, and provide an efficient and ergonomic workspace that conforms to the Human Factors requirements set out in the International Standard ISO 11064.

The control room will act as the building manger's "eyes and ears" making the design extremely important. When selecting a room, the architect/design team should consider the control room objectives and determine how much equipment and people the room will house. The size of the room should accommodate all necessary equipment, while allowing for people to comfortably move about.

A rectangular room shape provides the most options for equipment, display and console positioning. Rooms with sharply angled walls or with support columns should be avoided. To maintain the most consistent operator environment, a separate equipment room should be provided to house CPUs, servers, and other rack-mount equipment.

All functions related to safety and security will be performed from this location. These include issuing photo identification badges for staff and controlling keys or access cards for the facility.

Storage

General storage rooms should be located proximate to event spaces. Storage shall be provided for meeting room equipment including chairs, tables, risers and podiums. Storage capacity shall be limited to approximately 50% of the equipment inventory as it is assumed that at least 50% will always be in use. Meeting room storage shall be distributed throughout the back-of-house corridors proximate to the meeting rooms. This type of storage shall be in designated areas (striped floors), alcoves or otherwise securable areas.

The space provided is meant to ensure the security of the equipment while minimizing storage uses in circulation space, which would impair other facility functions. Doorways, fence gate openings and corridor geometry must be adequate to handle the maneuvering of chair and table carts, stages and food service trolleys.

Adequate number of large, secure storage rooms shall be provided proximate to the loading dock or the exhibition hall for the exclusive use of show decorators and exhibitors. Provide access for pallet jacks and forklifts via an overhead door, with a separate personnel door. These rooms shall be used for temporary storage of packing crates and other materials associated with displays while the event is in progress.

Durability of materials and security against intruder access are the primary design and operational concerns.

Client Recording Room

This room shall be designed to provide event managers a secure place to set up audio and video equipment to make distribution copies of an event. It should be located back-of-house and adjacent to the audiovisual/sound control room, with interconnecting conduit or cable trays.

Audio Visual/Sound Control Room

An audiovisual control room shall be centrally located back of house with other conference/exhibition hall support areas. This is the location from which event-related signals can be assigned throughout the facility. Satellite feeds and truck dock connections shall terminate here.

Chorus Dressing Room

One large dressing room shall be provided to support production-style events in the grand ballroom. The layout should anticipate that this room will be occasionally divided using pipe and drape or a rolling, nesting partition system to provide privacy by gender.

Access to this space shall be restricted to the back- of-house service corridors so that performers can be assembled for production cues. Quick access for costume changes is important. The dressing room does not have to be on the ballroom level, if convenient mezzanine space is available. Coordinate circulation patterns with food service staging areas to minimize traffic conflicts.

Green Room

This suite will serve as the before-and-after-event lounge for featured presenters. Each suite should connect via the back-of-house service corridor to the Grand Ballroom. The green room does not have to be on the same level as the ballroom but may be located on a mezzanine level above or below.

4.2.1 Health and Fitness Suites

Health & Fitness suites, incorporating fitness and exercise spaces, shall be provided to suit specific needs and market requirements. The Suites shall be part of retail establishment for use by registered members.

User demographics and facility requirements will impact upon the design and size of the facility. It is vital that the service provider identifies market criteria at an early stage. There are a wide range of potential disciplines.

The space should be flexible and able to accommodate new classes, programs or trends that may become popular in the future. A typical health and fitness suite will consist of: allow views of the main amenities, allowing members to orientate themselves as well as display the facilities.

- Reception Office
- Changing Rooms & Washrooms (including accessible facilities)
- Fitness Gym
- Studio(s)
- Plant Room
- Storage
- Staff Facilities appropriate to size of the facility

Split level health and fitness suites spread over more than one level, may utilize a feature stair located in the foyer/ reception. Lifts are required in any facility split over more than one level. These should be strategically placed to minimize horizontal travel, clearly signposted and easy to locate from the main entrance.

Accessibility

The Health and fitness suites should be designed in compliance with the Americans with Disabilities Act (ADA) and Disability Discrimination Act (DDA) Accessibility Guidelines for Buildings and Facilities. Examples of additional facilities:

- Café and/or juice bar
- Retail outlets
- Swimming, training or leisure pools
- Health spas, e.g. saunas, steam rooms and cold plunge pools
- Health and beauty treatments, e.g. massage, relaxation, alternative therapies, hairdressing and manicure
- Crèche
- Squash courts
- Tennis courts
- Physiotherapy/sports injury clinics
- First Aid room

Facility Planning

The fitness facilities forming part of a larger center will require a separate reception point with its own access control system. The fitness entrance should be clearly visible and enticing.

Fitness Centre Program Requirements Facilities Include:

- Male Fitness Centre
- Female Fitness Centre

Male Fitness Centre

General

The Male Fitness Centre should be positioned adjacent to the outdoor swimming pool, and allow convenient car access for members. A common entrance and reception area should control entry to the Male Fitness Centre and Outdoor Swimming Pool. The location of the Reception should ensure that all members and their guests entering the facility are obliged to register their arrival. Consequently, any other access to the area will need to be controlled, so that an outsider cannot avoid the Reception Entrance.

A retail space (Pro Shop) should be in the reception area, which will assist strong retail sales of sporting equipment and therapy products. The retail space should allow the members to touch and sample the displayed products, yet be secure out of operating hours.

A lounge area directly accessible from the reception desk is required. The lounge area should be a key area for socializing, and be a meeting point. The area should have tables and chairs, television and a standing juice bar. The lounge should be designed to accommodate a minimum of 20 persons, and should have a direct view into the gymnasium and an aerobic studio.

Fitness Centre

General

The focus of the Fitness Centre should be a combination of exercise equipment and such specialized facilities as steam rooms, plunge pools and saunas. The overall fitness gym area will depend upon the anticipated number of users and mix of equipment.

The minimum required space is $25m^2$.The shape of the fitness gym area will inevitably be defined by the overall design of the building, but ideally should aim to be broadly rectangular with a length to width ratio below 3:1. Adequate space allocation is important to ensure the required range of equipment and facilities accommodated.

Calculation of the total area and capacity of the fitness gym should be based upon a floor area of $5m^2$ per piece of equipment. This includes an allowance for circulation space around the equipment. The equipment mix shall depend upon the target market.

For general use, the split ratio of cardiovascular (CV) equipment to resistance equipment should be approximately between 40% to 60%, however this will depend on local need and demand.

Careful consideration is also needed for users with disabilities, including access for wheelchair users, which may impact upon space allocation and layout.

Table 4.2.1a Minimum Space Requirements		
Type of Space	Machine Footprint Range	Circulation
Resistance Area	2.0 m² per machine	1.75 x machine footprint
Cardio Vascular Area	1.5 m² -2.0 m² per machine	1.75 x machine footprint
Free Weights Area	2.5 m²-3.5 m² per machine	2 x machine footprint
Stretch Area	2.0 m² - 2.5 m² per person	

In addition, allowance should be made for a Fitness gym desk, information area and assessment rooms. The equipment supplier shall be consulted to obtain best advice on the space requirements around each piece of machinery. Table 4.2.1a provide guidance for minimum space requirements and include appropriate clear space to IFI accredited equipment as recommended by the Inclusive Fitness Initiative.

Circulation includes the area immediately the machine and the minimum aisle width to access the machine, but does not include the general circulation of the room.

Studios

The number of studios and the size of each studio required will be determined by:

- Number of simultaneous classes.
- Type and range of programs.
- Frequency and duration of each class.
- Number of attendees for each class.

It is vital that providers/operators identify the programs they intend to provide at an early stage. This affects both new facilities and any improvements or alterations to existing facilities. When designing a studio, the space may vary depending upon:

- The current market.
- Changing trends over time.
- The range of activities to be catered for.

Table 4.2.1b Multi-pur pose studios Minimum dimensions	
Execise Type	Dimensions (L x W x H)
Movements (Small Groups)	12 x 9.1 x 4.5 m
Rhythmic Gymnastics	14 x 14 x 9 – 15 m
Movement (Average Groups)	12 – 15 x 12 x 4.5 m
Movement (Large Groups)	21 – 24 x 12 x 6.1 m

Provision should anticipate peak time usage where possible, although this will inevitably lead to underutilization during other periods. It is common for fitness gym users to arrive at the facility already changed. It may therefore be possible to discount the number of changing spaces needed specifically for the fitness gym by between 25% - 35% of the number of workstations.

However, factors such as the facility's location, expected catchment area and brand success can all impact upon the changing room requirements, and the operator should be consulted at an early stage of the design to more accurately assess changing provision.

The studio should preferably be square, or rectangular with a length to width ratio of approximately 3:2. Instructors generally stand facing the users on the long side of the studio. Columns, projections and splayed walls should be avoided for safety.

Locker Rooms

Locker room capacities and sizes shall be calculated to meet the likely normal maximum occupancy level and patterns of use. Each facility will require an individual assessment of capacity and layout, however as a guide the following assumptions may be considered:

Fitness gym: Where the fitness area is relatively small, for example, when part of a small center one changing space should be provided for each item of equipment. For larger centers changing spaces and lockers shall be provided at a lower ratio, as gym usage is individual and users arrive and leave at different times.

Provision should anticipate peak time usage where possible, although this will inevitably lead to underutilization during other periods. It is common for fitness gym users to arrive at the facility already changed. It may therefore be possible to discount the number of changing spaces needed specifically for the fitness gym by between 25% - 35% of the number of workstations.

However, factors such as the facility's location, expected catchment area and brand success can all impact upon the changing room requirements, and the operator should be consulted at an early stage of the design to more accurately assess changing provision.

The changing rooms will also need to be able to cater for the demand of the participants of studio classes, particularly at peak periods as classes start and end. Capacities should therefore be based on the needs of large groups of people to use the changing rooms simultaneously, even after considering that several users may arrive changed or return home to shower and change:

- Allow one changing space for each $5m^2$ of studio floor area.
- Allow for 1.5 lockers for each person using the studio(s) over a one-hour period.
- Allow for one shower for every six changing spaces.

Where the fitness center includes a wet facility (e.g. cold plunge pool) the changing rooms should be designed to separate wet and dry foot traffic.

Changing rooms should typically consist of:

- Changing areas
- Toilets
- Shower areas
- A suitable range of lockers
- Vanity area with mirrors and hair dryers

Male Grooming Salon

Provide an area for male hair styling, and manicure and pedicure. Three hair styling stations and two Manicure / Pedicure client stations should be provided.

Female Fitness Centre

General

The Female Fitness Centre should be positioned adjacent to the indoor swimming pool, and allow convenient car access for registered members and their guests. A common entrance and reception area should control entry to the Female Fitness Centre and Indoor Swimming Pool.

The location of the Reception should ensure that all customers entering the facility are obliged to register their arrival. Consequently, any other access to the area will need to be controlled, so that an outsider cannot avoid the Reception Entrance.

A retail space should be located in the Reception area, which will assist strong retail sales of therapy products. The retail space should allow the guest to touch and sample the displayed products, yet be secure out of operating hours.

A lounge area directly accessible from the reception desk is required. The lounge area should be a key area for socializing and be a meeting point. The area should have tables and chairs, television and a standing juice bar. The lounge should be designed to accommodate a minimum of 20 persons.

Fitness Centre

General

The focus of the Fitness Centre should be a combination of exercise equipment and such specialized facilities as steam rooms, plunge pools and saunas.

Gymnasium

The overall fitness gym area will depend upon the anticipated number of users and mix of equipment. The minimum required space is 25m².The shape of the fitness gym area will inevitably be defined by the overall design of the building, but ideally should aim to be broadly rectangular with a length to width ratio below 3:1.

Adequate space allocation is important to ensure the required range of equipment and facilities that wii be accommodated. Calculation of the total area and capacity of the fitness gym should be based upon a floor area of 5m² per piece of equipment. This includes an allowance for circulation space around the equipment. The equipment mix shall depend upon the target market.

For general use, the split ratio of cardiovascular (CV) equipment to resistance equipment should be approximately between 40% to 60%, however this will depend on local need and demand.

Careful consideration is also needed for users with disabilities, including access for wheelchair users, which may impact upon space allocation and layout.

In addition, allowance should be made for a fitness gym desk, information area and assessment rooms.

The equipment supplier shall be consulted to obtain best advice on the space requirements around each piece of machinery. Table 4.2.1a provide guidance for minimum space requirements and include appropriate clear space to IFI accredited equipment as recommended by the Inclusive Fitness Initiative.

*** Circulation includes the area immediately around the machine and the minimum aisle width to access the machine, but does not include the general circulation of the room.**

Studios

The number of studios and the size of each studio required will be determined by:

- Number of simultaneous classes.
- Type and range of programs.
- Frequency and duration of each class.
- Number of attendees for each class.

It is vital that providers/operators identify the programs they intend to provide at an early stage. This affects both new facilities and any improvements or alterations to existing facilities.

When designing a studio, the space may vary depending upon:

- The current market.
- Changing trends over time.
- The range of activities to be catered for.

The studio should preferably be square, or rectangular with a length to width ratio of approximately 3:2. Instructors generally stand facing the users on the long side of the studio. Columns, projections and splayed walls should be avoided for safety.

Locker Rooms

Locker room capacities and sizes shall be calculated to meet the likely normal maximum occupancy level and patterns of use. Each facility will require an individual assessment of capacity and layout, however as a guide the following assumptions may be considered:

Fitness gym: Where the fitness area is relatively small, for example, when part of a small center one changing space should be provided for each item of equipment. For larger centers changing spaces and lockers shall be provided at a lower ratio, as gym usage is individual and users arrive and leave at different times.

Provision should anticipate peak time usage where possible, although this will inevitably lead to underutilization during other periods. It is common for fitness gym users to arrive at the facility already changed. It may therefore be possible to discount the number of changing spaces needed specifically for the fitness gym by between 25% - 35% of the number of workstations. However, factors such as the facility's location, expected catchment area and brand success can all impact upon the changing room requirements, and the operator should be consulted at an early stage of the design to more accurately assess changing provision. Allow for one shower for every six changing rooms.

Studio changing requirements

In addition to fitness gym members, the changing rooms will also need to be able to cater for the demand of the participants of studio classes, particularly at peak periods as classes start and end. Capacities should therefore be based on the needs of large groups of people to use the changing rooms simultaneously, even after considering that several users may arrive changed or return home to shower and change:

- Allow one changing space for each 5m² of studio floor area.
- Allow for 1.5 lockers for each person using the studio(s) over a one-hour period.
- Allow for one shower for every six changing spaces.

Where the fitness center includes a wet facility (e.g. cold plunge pool) the changing rooms should be designed to separate wet and dry foot traffic.

Changing rooms should typically consist of: Changing areas

- Toilets
- Shower areas
- A suitable range of lockers
- Vanity area with mirrors and hair dryers

Female Grooming Salon

A fully serviced lady's beauty salon which would provide hair washing and styling facilities, and manicure and pedicure stations should be provided. Four hair styling stations and three Manicure /Pedicure client stations should be provided as a minimum.

4.2.2 Swimming Pools

General

Outdoor pools for exercise and/or recreational swimming as appropriate for the market location should be positioned to receive direct sunlight from 11:00 a.m. until 5:30 or 6:00 p.m. throughout the season when the climate permits their use (except that in areas with extremely hot climates, where the direct rays of the sun are unbearable, the program may specify that the pool be partially shaded, and the water cooled).

Local customs require that the swimming pool not be in view of the public in general. In such cases, the program should require that the pool be screened from any view outside the immediate pool area.

Access to the outdoor swimming pool and pool terrace area should be so arranged that the public enter the area through a single passage which can be controlled by the fitness center reception. The separation between this area and the remaining grounds and terraces need not be more than symbolic, such as a low wall or hedge and a change of level, to indicate where free passage is not desired.

Access to the indoor swimming pool should be directly from the Female Health Club reception.

Occupant Capacity and Size of the Facility

The specialist Design professional need to consider the Theoretical Peak Occupancy (TPO) of the swimming pool facility as part of the design process. This will require calculation and integration of peak occupancy numbers for the water as well as the surrounding deck and seating areas. The rationale for the TPO density factor numbers for specific to the facility are: The density factor is established at 1.86 m^2 per person. This represents an average horizontal swimmer occupying a 1.5m by 1.2m area. Assuming a swimmer is swimming horizontally; a full body length is an average 1.5m with a 1.5m span to equal 2.3m^2.

There is a need to account for higher densities in shallow areas where swimmers wade vertically versus swim horizontally. Allow 1.9 m^2 per person for the shallow end. Assuming that the swimming pool is part of a complex that includes other outdoor facilities (such as Fitness Centre, tennis courts, and squash courts), following are the generally accepted criteria for estimating the number of swimmers: The maximum attendance of the recreational facility on the peak day can be estimated to be 68% of the total membership. Maximum attendance at the swimming pool facility can be estimated to be 40% of the projected maximum attendance on the peak day. The maximum number of swimmers is approximately 33% of maximum attendance.

Deck and visitor areas allow a density factor of 4.6m^2 per swimmer of deck space. When adding seating and tables, which separate groups, the square footage allows for a reduced density. For visitor areas, apply a density factor of 0.6m^2 per visitor. This seating is generally well above the water level.

Prior to applying the above guidelines, the social and economic conditions of a particular local community must be taken into account when designing a swimming pool facility. Swimming pool occupancy, or capacity, restrictions are subject to local regulations and vary from one jurisdiction to another.

General Physical Character

Before commencing the design,it is important to determine the style of pools the facility requires and the impact this will have on the space available for mechanical systems. The pool should be of an interesting, irregular shape, and of a minimum surface area equal to a rectangular pool 25 m X 15 m in size.

The shape should be such as to provide two parallel sides approximately 10 m long and 20 m to 25 m apart, which can be used for swimming back and forth for exercise. The maximum depth should be 2.5m and the minimum depth 1.0 m. The bottom of the pool should slope toward the main drain. Where the water depth is less than 1.5 m, the bottom slope should not exceed 1:12 horizontal. Where the water depth exceeds 1.5 m, the bottom slope should not exceed 1:3 horizontal.

Deck areas should accommodate chaise lounges and upright chairs and tables, the number of which should be based on facility size, market demand and the food and beverage concept. Provide a minimum 2.0 m deck around the perimeter of the pool with a total area of 0.25 m^2 per square meter of water surface.

Safety is of prime importance in the design of swimming pools and the surrounding areas. Pools and pool decks should be designed in accordance with the US National Spa and Pool Institute (NSPI) guidelines or equivalent standard, and all local codes and regulations.

Equipment locations should be established during the preliminary design phase. It must be decided, for example, whether equipment is to be in the pool structure or in a separate enclosure (keeping in mind that it is usually desirable to combine all of these facilities under a single enclosure).

The water filter assembly should be housed in an area with ample storage space. The filter equipment also should be in the filter room for easy and efficient operation and maintenance. Consideration needs to be given to the location of the pumps in relation to the water levels in the pools. Wherever possible, the pool pumps should be located below the water level determined by the gutter system or surge tank, so the pumps will have positive suction.

Accessibility

All outdoor/indoor swimming pool designs shall be compliant with ADA and DDA standards. The pool design shall not create SAFETY hazards with regards to maintaining necessary clearances, not infringing upon the recirculation of swimming pool water, or creating areas for potential entrapment.

4.2.3 Outdoor Tennis Courts

Tennis courts shall be planned in an open, unshaded area, with good natural drainage, and substantially a north-south orientation. For a single court, the minimum enclosed area shall be 18.3 m x 36.6 m.

Multiple courts within a single enclosure shall be separated by at least 3.65 m. Enclosing fences shall be 3.00 m high.

4.3.0 Commercial Offices

General

Entrances to upper story office uses shall be clearly distinguishable in form and location from retail entrances. Space planning for commercial offices require an understanding of the client requirements, the building envelope and location of the space being planned and the ability to conceptualize how people will use and interact within the finished built space – reflected on a two-dimensional plan.

The architect shall bring all concepts, both tangible and intangible, together in functional and creative layouts to optimize the space and floor plan within the building constraints. Prior to starting a space plan, the architect shall clearly understand the space to be planned, the culture, business, philosophy, needs and desires of the client for whom the planning is being done.

Some buildings are designed and constructed to house a client with tailor-made rooms and spaces to accommodate specific needs or desires. Other buildings are designed as speculative buildings to house a variety of clients and businesses.

The Facilities Standards shall be used in conjunction with the specific building program for each project, which delineates all project information, such as number and sizes of building spaces, and requirements for mechanical, electrical and other operating systems. It is imperative that each building be designed so that all components comprise an integrated solution, so that operation of the facility, energy usage and other criteria may be maximized.

4.3.1 Space Plan Components

A decisive factor for success in office design is selecting the right office type for each company (client).It is an important selection process in the context of a requirement analysis, because every office layout offers different solutions in terms of communication, concentration, flexibility and efficient use of space. The space planning process shall begin with the task of charting the organizational structure of the company, identifying personnel, their tasks, and necessary equipment, analyzing the operational process, identifying important sustainability factors, and gaining an understanding of the human and cultural qualities of the tenant organization.

Structural systems, construction materials, fenestration types, building shape and configuration, and architectural design and detail have a major effect on space planning decision. Architects and structural engineers shall be in close consultation to determine the degree of flexibility and openness that can be created; The length of a building's structural spans (bay size) would be of essence, as small bays will restrict partition and furniture placement and limit traffic flow.

A combination of structural system and materials factors will govern, where, and how door and window openings can be made in both interior and exterior walls. The relative simplicity or complexity of a building's shape or configuration will determine its usefulness for a given set of planning requirements. In space planning terms, the design program shall be a written document that qualify and quantify the client's or the user's needs for a given project. Additionally, the design program shall be accompanied by relationship diagrams that express physical planning relationships accurately.

All interior spaces and space plans regardless of nature or size of business, are made up of the same basic components. These include an entry and reception area, typical rooms, semi-typical rooms, food rooms, support rooms, staff work areas, furniture and equipment, circulation or corridors, building core, accessible areas or rooms, and building structure and infrastructure.

For pur poses of construction and obtaining b uilding permits, the IBC occupancy classification "Business Group B" shall be applied.

4.3.2 Work Area Mix Criteria Matrix

The architect shall develop a Criteria Matrix (CM) that will visually organize information of a variety of factors of the design program in a condensed form. The CM shall include daylighting and shading requirements for some of the spaces in the column titled "Sustainable Factors." A sample format is shown in Figure 4.

A factor of 25% to 35% of the total square meterage calculated in the CM shall be added as an estimate for circulation and partitions. This figure will vary from project to project, depending on the configuration and construction of the building shell and the nature of the functions to be performed in the space in general terms. When functional planning requirements demand many separate spaces, the circulation and partition factor will be higher than normal.

The mix of offices and workstations in some corporate settings have fewer offices and more workstations. The space planning analysis shall focus on the concepts of zoning and isolation. When user and spatial needs are being organized, as in the development of a CM, acoustic needs for privacy, isolation, and absorption shall be identified. The architect should create quiet and noisy zones (often coinciding with private and public zones) during the bubble diagramming or block planning phase.

Acoustic conflicts shall be addressed through appropriate and sensitive space allocations. Where acoustic conflicts cannot be resolved easily through space planning, the acoustic interference shall be dealt with through means other than zoning and isolation. The specialist acoustic consultant with knowledge of the sound levels generated and the construction techniques needed to limit transmission to acceptable level, shall guide the architect in resolving acoustic conflicts. Multiple uses of space are common in Mixed-Use developments, and high acoustic ratings are essential.

In general, each office requires two to three times as much space as an individual workstation. The architect shall present the advantages and disadvantages to the Owner for review and comment.

Any one of the two basic methods shall be used to establish the average square meter (m^2) per person: Industry accepted standard or actual square meterage calculated during the programming phase.

Depending on the type of business, most businesses average between 17.65m^2 and 20.90m^2 per person for typical office space. The average square meterage per person will include the actual work area for each employee, a portion of all other rooms and circulation within the office space.

Planning Rules of Thumb Table

Table 4.3 Rules of Thumb	
Space/Function	Space Requirement
Reception	2.8 m^2 per person
Conference Room	2.3 – 3.25 m^2 per person
Private Offices (standard work and consulting space)	11.0 – 14.0 m^2

Table 4.3 Rules of Thumb	
Private Offices (Executive Office w/lounge seating)	18.5 – 28.0 m^2
Open Plan Office Space	5.25 - 7.0 m^2 per person
Prayer Room - Male	20 m^2 minimum
Prayer Room - Female	20 m^2 minimum
Ablution Area	20 m^2 minimum

The preliminary space planning process shall produce the approximate size of typical rooms (spaces). Table 4.3 provides a preliminary list of frequently used ces and functions with their approximate space requirements.

4.4.0 Residential Apartments

4.4.1 General

The primary objective in planning residential apartments is to provide a desirable place to live for the intended end-users. Different parameters need to be applied and different elements are selected depending on who the end-user is to be.

Buildings where people live and sleep are designated as Group R occupancies. The Residential facilities shall be classified under Occupancy Group R-2, as the occupants are primarily permanent residents in an apartment.

Vertically mixed-use buildings shall generally be designed with commercial storefronts on the ground floor and residential uses above. Entrances to upper story residential uses shall be clearly distinguishable in form and location from retail entrances. The mix & size standards which are given in the following paragraphs of this section are based on established norms for a standard Residential Apartment category building in metropolitan capital cities or other major centers with well-developed mixed economies, and full support facilities. They represent a well-balanced operational and economic relationship between the various apartments, public area, back of house and external facilities which make up the apartment complex of an up-to-date building of its class.

The standards serve, therefore, as the essential point for establishing the basic design criteria for a prospective new building; or as a yardstick for measuring the adequacy or otherwise of established buildings, or when contemplating major design changes as part of a rehabilitation program. However, they must be subject to critical evaluation in the light of the differing circumstances which inevitably confront each building. Each will have its own special needs, dictated by its economic environment, site planning constraints, local traditions, labor practices, support facilities and market demands.

For example, the building may not be typical of the standard model with its full support facilities including competing local residential occupancies. In some cases, a building may be in an area with limited near-by full support facilities. Where a building may be in a remote or developing area with virtually no support facilities in the same class. Such variations in environmental and economic factors must affect the planning criteria. There are influences which, revealed by initial planning and feasibility studies, may warrant variation of one or more specific standards.

Figure 4.0a Male Fitness Center Functional Relationships

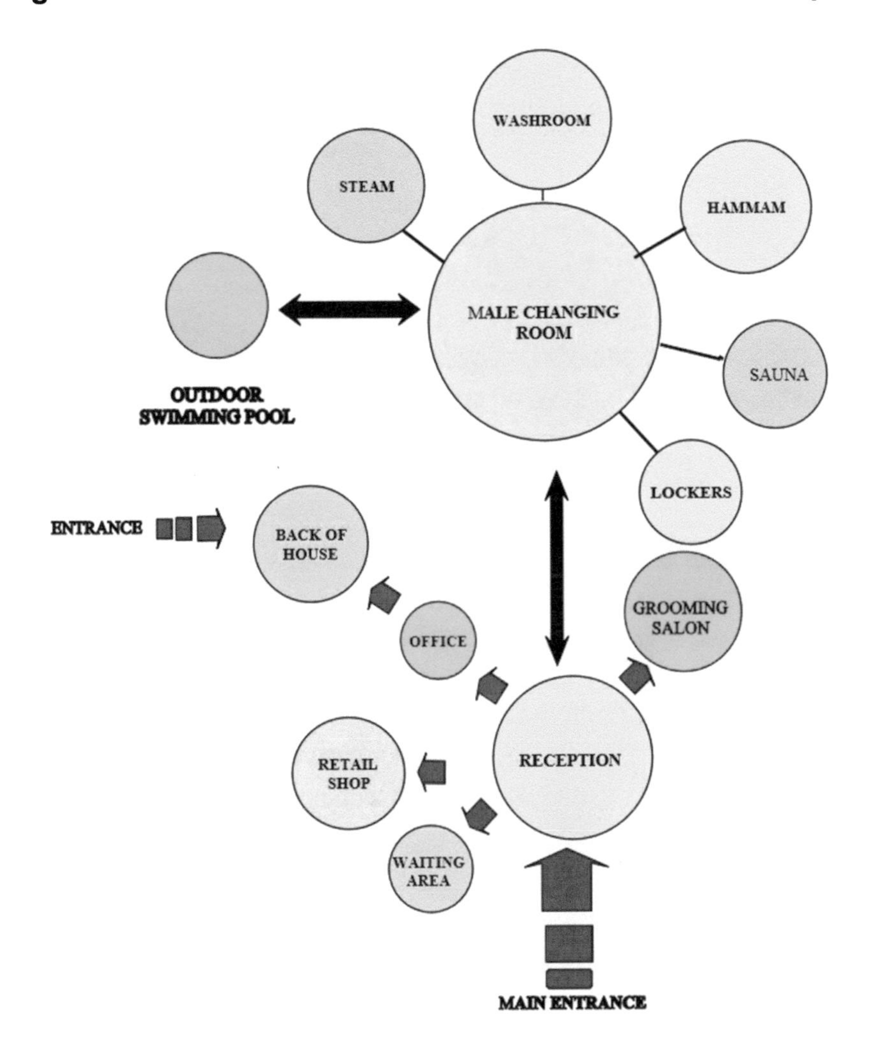

4.4.2 Space Plan Components

For people to have a habitable space in which to live, the IBC requires minimum dimensions in a building. Ceilings in habitable spaces must be a minimum of 2.3 m to provide people sufficient headroom to feel comfortable.

A habitable space is defined as a space in a building for living, sleeping, eating, or cooking. Bathrooms, washrooms, closets, storage or utility spaces, and similar areas are not considered habitable spaces. The IBC also requires habitable rooms to be a minimum of 2.13 m in width. This provides adequate room for furniture and space for the intended activity. The only exception to this rule is non-accessible kitchens for which a minimum clear passageway of not less than 910 mm must be provided. This provision is typically applied to residential occupancies and is consistent with the requirements in the International Residential Code.

Residential dwelling units are also required to have at least one room that has a minimum of 11.15 m². All other habitable rooms are required to be a minimum of 6.5 m² in area.

All standards relating to fire emergency escape measures referred to in this section are based on the building being fully sprinklered. Fire emergency exits shall be located so that no apartment or service room entrance door is more than 30 m. from the emergency exit door. Nor shall any such room be in a "dead end" extension of the main corridor where its entrance door is further than 10.7 m. from the nearest fire emergency exit door on the same corridor. No other "dead end" or secondary corridor opening off the main corridor or lift lobby opening off a main corridor, shall be longer than 10.7m.

All partitions shall extend from the structural floor slab to the structural ceiling slab on each floor. All pipes and ducts penetrating partitions shall be sleeved and sealed to preserve acoustic and fire protection properties of the partitions. All electric boxes shall be offset from each other between adjoining rooms for the same purpose. Partitions between public corridors and apartments, and between bedrooms are designated as fire rated walls and shall have one-hour fire resistance capability.

Unless view is a primary consideration, apartment wings shall be oriented so that apartments face north and south on open sites.

Guest lifts and the service core shall be centrally located in relation to the residential apartment layout. The service core shall include a service stair which connects every level of the building.

All living spaces within the apartments shall be provided with direct natural light without exception. Natural light shall be provided to apartment floor public corridors wherever practical.

4.4.3 Space Planning Concepts

Entry Canopy & Vestibule

The primary entrance drive shall provide two-way circulation through a lighted and landscaped corridor to guest parking and the primary apartment entrance. Two- way main entrance driveways should have a minimum width of 9 m and inside turning radius of 10 m. A minimum of one of the traffic lanes (minimum 2.5 car widths), in front of the main entrance shall be covered by a canopy as protection against precipitation. This canopy shall extend over a minimum of 7.6 m in length at the main entrance and shall cover also the loading platform from the kerb to the apartment lobby. This canopy shall have a clear height of 4 m or more if required by local regulations.

The outer traffic lane shall remain uncovered for passage of higher vehicles, such as fire fighting vehicles, etc.

The apartment complex entry should convey an inviting message, and provide an initial orientation by welcoming the tenants/visitors and showcasing the building or identity. Entrance walls must be largely transparent (glazed). Doors should convey indigenous style and character. Vestibules shall provide convenient entry by tenants/visitors without requiring the need to open doors and shall provide a separation of interior and exterior climates.

Typical Apartment Floors

All standards relating to fire emergency escape measures referred to in this section are based on the building being fully sprinklered.

All doors in partitions shall have a minimum 20-minute fire resistance capability. Where any duct penetrates a fire rated wall, a fire damper shall be provided with accessible means of re-setting the damper.

Apartment floors shall be structurally conceived to ensure that the center partition between each pair of apartment modules is non-structural and capable of being penetrated, removed or otherwise adapted to meet changing rental needs.

All vertical shafts through the apartment floors, including all engineering services between apartments, shall be tightly sealed at the floor level of the lowest apartment floor; at the ceiling of the of the highest apartment floor, and at such intermediate levels as required by Chapter 7 of the current version of the IBC and NFPA 101.

Generally, the following planning requirements shall be met in respect of a new building on each floor:

- Construction bays shall be designed to a cost effective minimum module.
- Corridors shall be not less than 1.8 m, provided corridors are opened to a width of 2.0m at paired apartment entry doors.
- A minimum of two emergency escape stairs, (one at each end of each corridor)– 16m² each.
- One service stair to be located adjacent to the service lift lobby – 16 m².
- One linen room per apartment floor. The linen room shall open off the service lift lobby.
- Housekeeping service area, including supplies shelves, cart and equipment storage, staff toilet and cleaner's closet - 20 m².
- A room each for electrical panels and IT backboards for each floor.
- Service lift lobby - allow 2 m. in front of lift doors to maneuver service carts.
- Lift lobby shall be sized to accommodate waiting, and located to achieve noise isolation; finishes shall be consistent with apartment access corridor. The ground floor landing should be adjacent to the entrance lobby, and ideally within view of the reception/security desk.

Special Requirements

Provide a 800 mm diameter metal, vertical, gravity-type chute that complies with the latest version of NFPA 82, for transferring trash by gravity to a room at the bottom or to an interface to the compactor. Locate the chute between the housekeeping linen room and the service lift.

Provide a 600-mm diameter metal, vertical, gravity-type chute that complies with the latest version of NFPA 82 for transferring soiled linen by gravity to a room at the bottom or to an interface to the central laundry. Locate the chute between the housekeeping linen room and the service lift.

Residential Apartments

Although there are many factors to consider when defining the floor plan - that is, the way the apartments are arranged and the spaces within the apartment are used - is critical to the overall satisfaction of the residents. Floor plans can be evaluated in several ways. The following are some of the most important issues to consider when evaluating the spaces within a home.

I. *Zoning*

There are three basic areas or zones within the apartment. They include:

- The public zone, composed of spaces where non-family members (visitors) are generally allowed.

- The work zone, composed of areas where work activities that support the lifestyle of the household take place.
- The private zone. Composed of rooms used exclusively by the household members for private activities.

II. *Circulation*

Circulation paths within the apartment should be as short and direct as possible. A circulation pattern through the home that is efficient and that utilizes minimal floor space should be the goal of the architect. A central entryway and hallway often assist in the creation of a good circulation pattern in the apartment. Zoning and circulation complement each other in apartment design. A floor plan that exhibits good zoning also will have excellent circulation, and vice versa.

The central entryway shall lead to a hallway that opens to each zone in the apartment. The space devoted to this circulation path through the home will generally amount to approximately 5% of the total square meterage – a good circulation-to-usable-space ratio. Additionally, it allows circulation from one zone to another without violating the third zone.

The architect shall strive for efficient, fluid and discrete paths that allow multiple furniture configurations. In addition to hallway placement, the architect must also plan the circulation within rooms. The rooms within the apartment must also accommodate circulation throughout the apartment. A traffic pattern that cuts directly through a living room or family room is extremely undesirable. Entrances to rooms should be located so that traffic going through those rooms can move along one wall.

The placement of furniture should direct traffic flow in an open floor plan. Similarly, the architect must ensure to place bedroom closets near the entryway to the bedroom, thus avoiding a circulation path that cuts through the middle of the bedroom.

Organization of Space and Relevant Structural Components For maximum efficiency in a floor plan, the following rules should be considered when putting together space in the apartment:

III. *Orientation:*

Some aspects of orientation that should be considered when designing housing include: An east-west orientation is preferable to a north-south orientation in most locations. With an east-west orientation, which has the longer sides of the apartment facing north and south, windows can be placed on the south side of the apartment to take advantage of the sun's position (low on the southern horizon) during the cooler months of the year. These windows allow for passive solar heating during the cooler months as well as natural lighting during the day. By putting few or no windows on the east and west sides of the apartment, intense morning heat and light from the sun as it rises in the east and sets in the west can be avoided.

Windows can be positioned so as to take advantage of cross breezes. Placing doors and windows so that prevailing winds during cooler weather, can travel from one side of the apartment to another allows for natural cooling and ventilation.

IV. *Entryways*

How one enters and exits the apartment will have a direct impact on how the spaces are arranged in the residential unit. Entryways should be easily accessible for household members and their guests, but they should be secure and serve as a buffer between the exterior and the rest of the apartment. The following are some ideas to include in the design of entrances to the apartment:

- The front entry should be somewhat separated from the rest of the living areas in the apartment. A washroom for guests should be close to the front entry and other public areas in the apartment. However, it should not be situated so that it is the first room a guest sees upon entering the apartment.
- A service entry is an additional entry to the apartment, generally through a corridor leading from a service lift landing. In many cases, it is the entrance most often used by the household and others who provide services to maintain the apartment. The service entry should be convenient to the kitchen, so that household members can easily move groceries into their proper storage areas.

V. ***Work Areas***

Kitchen: The kitchen is the primary work area in the apartment. Kitchens must be designed to accommodate two or more persons, an array of household appliances, and a variety of activities that take place in this room in addition to meal preparation. There are four basic steps to planning an efficient kitchen:

- Plan the kitchen location and arrangement so that it is a convenient work space and is easily accessible from the other zones in the apartment, as well as from the service entry.
- Decide on a kitchen configuration that will minimize traffic flow through the work area. There are three basic shapes:
  - Corridor or two wall (Gallery Kitchen)
  - L – Shaped Kitchen
  - U – Shaped Kitchen

Choose suitable space standards: Allow for adequate counter space and storage near major appliances and by the sink.

Plan for efficient operation: Locate the appliances, storage, and counters in the place allocated for the kitchen.

Evaluate the configuration in terms of efficiency and ease of maintenance: The efficiency of kitchen spaces is often evaluated by how the space is used. Following are the essential zones:

Cooking: The cooking zone shall encompass the cooktop (and ovens if not part of the cooktop), or range, vent hood, a microwave and/or steam oven, and storage for cooking gear.

Preparation/clean-up: This area shall incorporate the primary sink; dishwasher; dish, glass and utensil storage; counter space for meal preparation; trash and recycling receptacle; and storage for prep and clean-up material.

Food Storage: This zone shall include the primary refrigerator and dry food pantry.

The following are potentially specialty zones:

Children Friendly: Many families would like to have spaces for their children to safely use the kitchen. With a beverage refrigerator and refrigerated drawers; storage of children’s snacks; dishes and cups; and counter or chair height seating for homework.

Space for Smart Devices: Provide hidden charging stations to store devices like phones and tablets so that they are always operable, at hand and charged.

Homework: A zone used for managing bills, shopping lists etc. should include space for a laptop computer, power outlet and adequate lighting. Provide a writing surface where meals can be planned, bills paid, and correspondence written.

Laundry: In a one-bedroom apartment, the placement of the laundry equipment shall be near the kitchen without being in the food preparation area. This configuration allows for maximum work efficiency and the dovetailing of tasks without the risk of contaminating food preparation surfaces with the harmful chemicals associated with laundry detergents, bleaches, and fabric softeners. The laundry equipment could be placed in a small room between the service entry and the kitchen or it could be in a deep closet within the kitchen that is not too close to the actual food preparation area.

In the two and three-bedroom apartments locate the laundry area near the bedrooms where clean clothes are stored and dirty clothing is most generally deposited after being worn. A laundry room near the bedrooms eliminates the need to carry baskets of clothing from the work zone to the private zone. A good option would be to locate the laundry area between the kitchen and bedrooms, so that it borders both the work and private areas. Wherever the laundry area is in the apartment, the most important consideration is that it be convenient for the user during normal routines.

In the two-level Penthouse configuration, locate a Laundry Room at second floor complete with one (each) washing machine and dryer, utility sink and clothes rod, ironing station, folding area and clothes drop.

Storage: A minimum of 10% of the apartment footprint shall be provided as storage space. Bedroom closets should be located near the bedroom entrance. Walk-in closets shall be provided in the two-level penthouse master bedroom. Standard wall closets can also be situated to serve as buffers between the private zone and the other zones in the apartment. This arrangement could serve to filter noise away from and distractions that might come from the public and work areas of the home so that household members trying to sleep or study in the private zone will not be disturbed.

Plumbing: In the two-level penthouse, plumbing should be concentrated by situating the kitchen, laundry area, and bathrooms next to each other or stacked on top of one another. In addition, these rooms need to be near the hot water generator.

***Privacy*:** With open floor plans, privacy is still a highly valued aspect. Cultural mores still uphold the need for private spaces in the home to carry out certain activities. The great room concept where living, family, and dining room are combined into one space have opened the public areas of the home, but the private zone is still a desirable and important design element. The private zone needs to be secluded and not in direct sight of the public area. visitors in the home should not be able to readily enter the private zone or even look in to it.

The private zone includes a hallway that leads to each household member's bedroom and to the bathrooms. The entrance to a private bathroom or bedroom should not be situated at the end of that hallway. It is essential to locate doors to private rooms off the sides of the halls so guests do not have a clear view of the interiors of those rooms.

Wheelchair Access: Appropriate space planning is critical for an individual who uses a wheelchair. Access to a door, through a door, and closing a door all require specific placement and dimensions. Once the individual is through the door, he/she must be able to use the fixtures and equipment within the space just entered. This design concept is particularly important in bathrooms and kitchens. ADA and DDA specific information on space planning for the disabled shall be included in the design of the apartment.

Facilities Included: The residential unit mix shall be achieved across the building by balancing the number of story's utilizing differing floor plates. The resultant apartment mix shall consist of the following:

- One-bedroom Serviced Apartments.
- Two-bedroom Serviced Apartments.
- Unfurnished two-bedroom Apartments.
- Unfurnished three-bedroom Apartments.
- Two level Penthouses – 4 bedrooms.

Notes

- Check local authority regulations for compliance.
- The areas given are based on normal needs for developed areas. Local market studies may show need for adjustments to the allocations in the tables.
- The two/three-bedroom units and the two-level penthouse shall have separate quarters for live-in help, labelled a maid's room. The maid's room with attached bath shall be located off of the kitchen. The maid's room shall have a minimum foot print of 6.5 m^2.

4.5.0 Back of House Areas

4.5.1 Storage

The facilities shall include the following:

- General or "Bulk" storeroom.
- Tenant storage.
- Miscellaneous Storage (linen, chemicals etc.).

Storage areas are to be located to ensure security, control of materials and operating efficiency through ease and speed of service with minimum staff requirements. Storage spaces for unissued materials should be grouped within an easily controlled area, with direct, fully supervised line of transport to point of ultimate use.

The Architect shall identify adequate space for long term tenant storage, preferably in unused basement space. Allow for10% of total individual apartment square meterage per apartment.

4.5.2 Housekeeping Department

The housekeeping department shall be located adjacent to the laundry sorting, and to receive clean linen directly, and should provide easy access to service lifts, and be as convenient as possible for issue of uniforms to employees.

Layout of the department, including storage shelving, uniforms storage, serving area, and office space should be made by the architect.

4.5.3 Laundry

As a source of noise and vibration, avoid locating laundry under or adjacent to principal public areas. If such proximity is unavoidable, special measures should be taken to eliminate any possible transmission of objectionable sound or vibration to such tenant/visitor areas. In case of an upper floor laundry installation, all washer-extractors should be suspension mounted. No rigid mounted washer-extractor should be installed on an upper floor.

It is imperative, particularly when dealing with composite floor systems, that the Structural Engineer checks the static, live load and dynamic force characteristics of the wash equipment. The Building shall be designed to facilitate off-premises laundry processing, and in-house valet facilities planned for the provision of 7 day a week/24 hour pressing and tenant laundry.

4.5.4 Employee Facilities

The number and size of the various locker and wash rooms required should be specified in the program. However, these requirements are based on conditions existing at the time of the survey, and locker rooms should be planned in a flexible manner, so that the capacities of adjacent locker rooms can be easily adjusted to accommodate changing labor conditions.

The locker rooms shall be divided into different sections, each with its own washroom and shower facilities. The divisions may include:

- Male Locker Room.
- Female Locker Room.

The entry to each locker room area shall be through a vestibule, arranged to provide effective screening from adjoining areas. Washrooms shall be so located that they can be entered directly from shower and locker rooms, and can be entered also without passing through the locker area.

Cleaner's closets shall be conveniently located, and the closet should not open off a toilet or locker area.

Provide employee lounge facilities closer to the locker rooms. Provide two separate employees' dining rooms (male and female), each divided into two or more bays. Local custom may require that supervisory personnel are served at tables in a separate Staff Dining Room. These considerations may result in the requirement of more than a single kitchen or service unit.

4.5.5 Maintenance and Engineering

This area should include engineering administration offices,workshop facilities for repair and activities required to maintain the apartment complex. The Engineers Offices should be arranged so that the Engineering Manager can give as complete and continuous a visual supervision practically to the entire Engineering Department.

In addition to the offices, the following facilities shall be provided:

- Carpentry/Upholstery Workshop.
- Electrical Workshop.
- Mechanical/Plumbing Workshop.
- Paint Shop.
- Engineering Store

4.5.6 Back of House Corridors

Service corridors provide circulation to back of house spaces and provide access to public areas for employees without interference with tenants or visitors. Service corridors should be minimum 1.80 m wide on trolley routes, and 1.4 m on other routes. Secondary corridors to be 1.2 m wide minimum.

A separate street entrance should be provided ensuring a secure and direct entry to the back-of-house for all staff.

4.5.7 Receiving Area/ Loading Dock

Provide an independent Service Entrance off a back of the building or side street, with a truck dock, or space for trucks to stand off the public street while making deliveries. Comply with requirements of local traffic authority. See section 2.1.9 for further details.

4.6.0 Major Equipment Spaces

General

These spaces include, but are not limited to, mechanical and electrical equipment rooms, enclosed cooling towers, fuel rooms, lift machine rooms and penthouses, electrical closets, telephone frame rooms, transformer vaults, incinerator rooms, and shafts and stacks.

Mechanical and electrical equipment rooms must be designed with adequate aisle space and clearances around equipment to accommodate maintenance and replacement, as recommended by the manufacturer and in compliance with local code requirements. The Mechanical and Electrical design engineers should be cognizant of the necessity to provide for the replacement of major equipment over the life of the building, and should insure that provisions are made to remove and replace, without damage to the structure, the largest and heaviest component that cannot be further broken down.

Mechanical equipment rooms should not be less than 3700 mm clear in height

4.6.1 Chilled Water Plant

Since the basic qualities of HVAC systems can vary greatly from building to building, broad generalizations are difficult to draw concerning the relationship between the space planning process and HVAC systems. As these systems are to be designed for maximum flexibility, and permit quick and easy changes, they have little impact on the space planning process. More specifically, the air distribution system shall be designed so that new partitions can be located with almost complete freedom in terms of HVAC requirements. A minimum of 1% of the building's gross area shall be provided for the central cooling, and domestic hot water heating plant (location to be agreed upon during preparation of concept submission).

Space requirements of mechanical and electrical equipment rooms shall be based upon the layout of required equipment drawn to scale within each room.

For operational flexibility of the refrigeration plant, allow space for a chiller of same capacity for future use. In larger buildings, a long, narrow room is usually preferable to a square one. The ceiling height at a chilled water plant varies from a minimum of 3.7 m for a building of moderate size to a maximum of 4.88 m for a very large building.

Chillers shall be placed to permit pulling of tubes from all units. The clearance shall equal the length of the tubes plus 600 mm. The size and number of cooling towers are related to the cooling requirements of the building. Cooling towers may be located on the ground if they are at least 30 m from any building or parking lot to avoid property damage and unhealthful conditions from the splash, fog, and microorganisms given off by the towers. An alternate location is the roof of the building. The towers should be isolated acoustically from the frame of the building; from the noise and vibration they generate. Noise-sensitive areas such as auditoriums and meeting rooms should not be located directly below them. Rooftop cooling towers must be located well away from windows and fresh air louvers.

4.6.2 Fan Rooms

The air handling equipment room shall be located centrally to reduce distances conditioned air must travel from the equipment room to the farthest air-conditioned space. A minimum of 4% of the typical floor's gross floor area shall be provided on each floor for air-handling equipment.

To facilitate the building to be zoned for better local control, multiple fans distributed throughout the building shall be planned. To minimize the vertical runs of ductwork, the designer shall aim to have a separate fan room for each floor of the building. Noise-sensitive areas such as meeting rooms, auditoriums etc. shall not be located adjacent to fan rooms. Air-handling units require a minimum clearance of 750 mm on all sides, except the side where filters and coils are accessed. The clearance on that side should equal the length of the coils plus 600 mm.

4.6.3 Domestic Water Pumps

Where the water service enters the building, a room is required to house the water meter and the sprinkler and standpipe valves. In a building, taller than three or four stories, a suitably sized suction/storage tank and a constant volume variable speed pumping station is needed to boost the water pressure in the domestic water system. A set of pumps are required for a sprinkler/ standpipe system. A heat exchanger to heat domestic hot water is often located in the same area.

4.6.4 Electrical Service Entrance, Transformers, Switchgear, and Emergency Power Supply

The locations and sizes of these elements vary considerably, depending on the size and purpose of the building, the type of electric service provided by the local utility supply company, the standards and practices of the utility company, the preferences of the building owner, the judgment of the electrical design engineer, and local electrical codes.

Primary transformers may be located either outside or inside the building. where the design team finds outdoor placement objectionable, the primary transformers shall be located within the building. Dry type primary transformers shall be in the main electric switch room, which is a fire-rated enclosure with two exits. The main electric switch room is usually located in the basement of the building.

The emergency generator required to furnish electricity during power outages, shall be located adjacent to the main switchgear room.

4.6.5 Elevators (Lifts)

Lifts serving the same zone of a building should be arranged in a single bank so that waiting persons can keep all the doors in sight one time. A bank of three in a row is the largest that is desirable; four in a row is acceptable. Passenger Lifts shall be grouped in banks of at least two for efficiency. Lift groups of four or more shall be separated into two banks opposite each other for maximum efficiency in passenger loading and minimum hall call notification for accessibility under requirements of ADA/DDA.

Escalators and Lifts shall be carefully sited to invite shoppers onto other levels. Lift shafts are noisy and should not be located next to occupied spaces.

4.6.6 Service Cores

Spaces for the vertical distribution of mechanical and electrical services in a building shall to be planned simultaneously with other building elements that are vertically continuous or that tend to occur in stacks— principally the structural columns, bearing walls, shear walls etc.; exit stair-ways; lifts and lift lobbies; In high- rise office buildings, where a maximum amount of unobstructed, rentable area is a major criterion for floor layout, a single central core is most logical. Vertical distribution shafts need to connect directly with the major equipment spaces that feed them and the horizontal distribution lines they serve.

The boiler room, chilled water plant, central fan room, exhaust fans, water pumps, sewage ejector, and cooling towers should cluster closely around the vertical distribution shafts. The electric switchgear shall be located closer to this cluster.

The electrical, telecommunications, and technology closets must stack up along the wiring shafts at each floor. The washrooms, and janitor closets shall back up to plumbing walls. Horizontal supply and return ducts need to join easily with the vertical ducts in the shafts, and horizontal piping for hot and chilled water distribution must branch off conveniently from the riser pipes.

4.6.7 Swimming Pool Plant Rooms

The plant room should ideally be positioned at one end of the pool, preferably at the deep end to ensure both hydraulic efficiency of the pool water circulation system and to minimize the length of the suction pipework. Location will have an impact on the size of the flow and return pipework as, depending on whether the plant room is below or above the pool water, this will affect the net pressure suction head (NPSH) or, less formally, the available pressure at the pump suction inlet, or head resistance.

This can also mean that pumping sumps for overflow drainage and forced draft ventilation may be necessary. The level of the plant room floor should be the same level as the pool static water line or below (which is the level of the water when at rest) but, should the plant room floor have to be located above the pool water level, then the vertical lift of the circulation pump/pumps should not exceed the pump manufacturer's recommendations.

There should be easy access from the poolside into the plant room for periodic monitoring of all the plant and equipment. The size of the plant room is critical. It should not only be large enough for the plant and equipment but also be big enough to meet future servicing, maintenance and replacement requirements. Clear working space around all the equipment is an absolute necessity. A storage space for chemicals separated from the rest of the plant room shall be provided with easy access for the delivery of the chemicals required.

Fig 4.3 Commercial Offices Functional Relationships

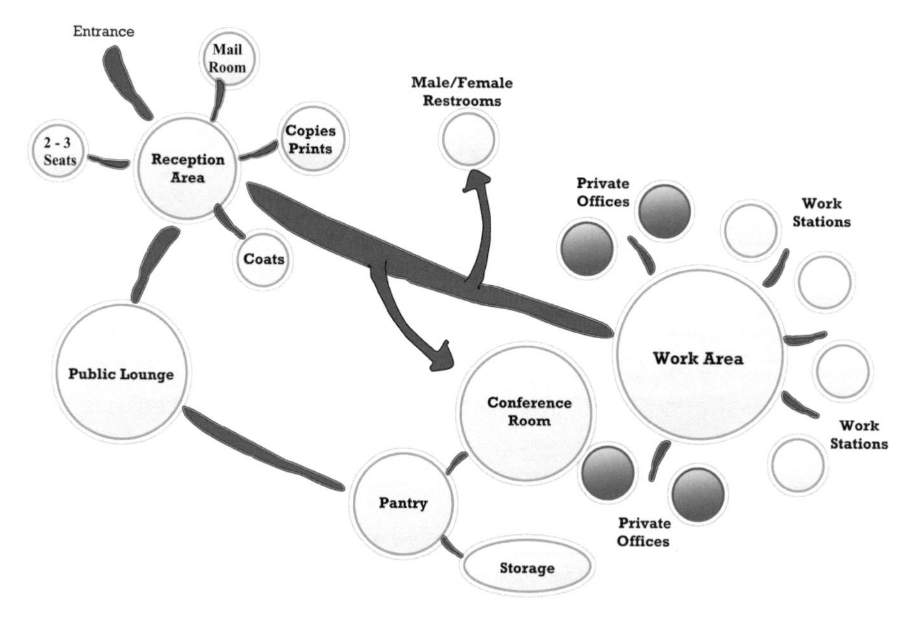

Figure 4.3 Criteria Matrix

CRITERIA MATRIX OFFICE BLOCK RELATIONSHIP	Square Meter Needs	Adjacencies	Public Access	Daylightand/ or view	Privacy	Plumbing	SpecialEquipment	SPECIAL CONSIDERATIONS
(1) Lobby/Reception Office	3	10	H	Y	N	N	N	
(2) Conference Rooms	1	6	M	I	Y	N	Y	
(3) Open Workspace	1	10	N	Y	L	N	Y	
(4) Team Space	3	10	N	N	N	N	Y	
(5) Cubicles	6	8	M	N	L	N	N	
(6) Private Offices	2	5	M	I	Y	N	I	
(7) Shared Office		8	N	N	L	N	N	
(8) Copier Room	3	5	N	N	N	N	Y	
(9) Storage Space		1	N	N	N	N	N	
(10) Pantry	3	2	N	N	N	Y	Y	
(11) Break Room		10	N	Y	N	Y	N	
(12) Restrooms			M	N	Y	Y	N	

LEGEND:
H = High
M = Medium
L = Low
Y = Yes
N = No/None
I = Important but Not required

Fig. 4.0b Female Fitness Cente Functional Relationships

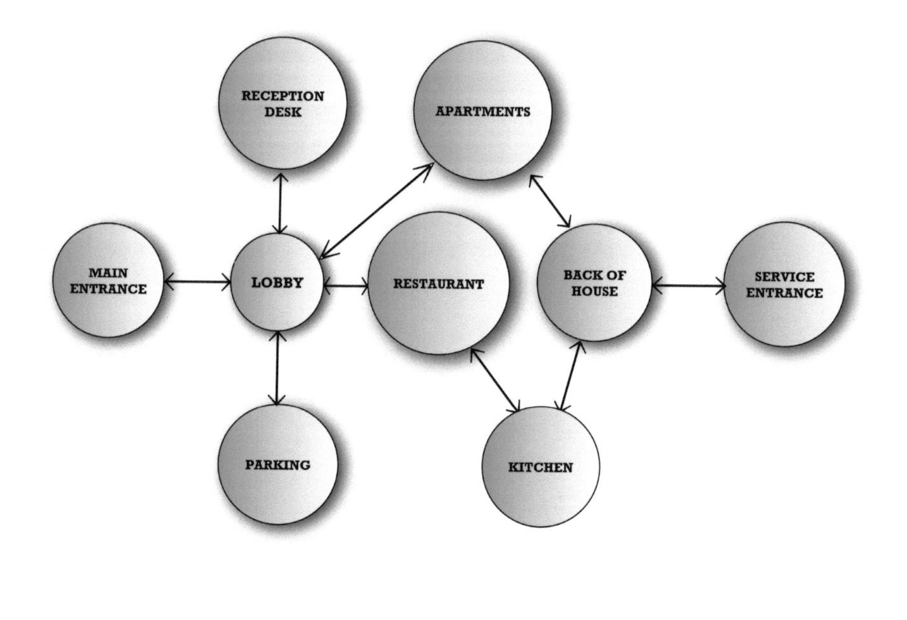

Fig. 4.4 Apartment Block Functional Relationship

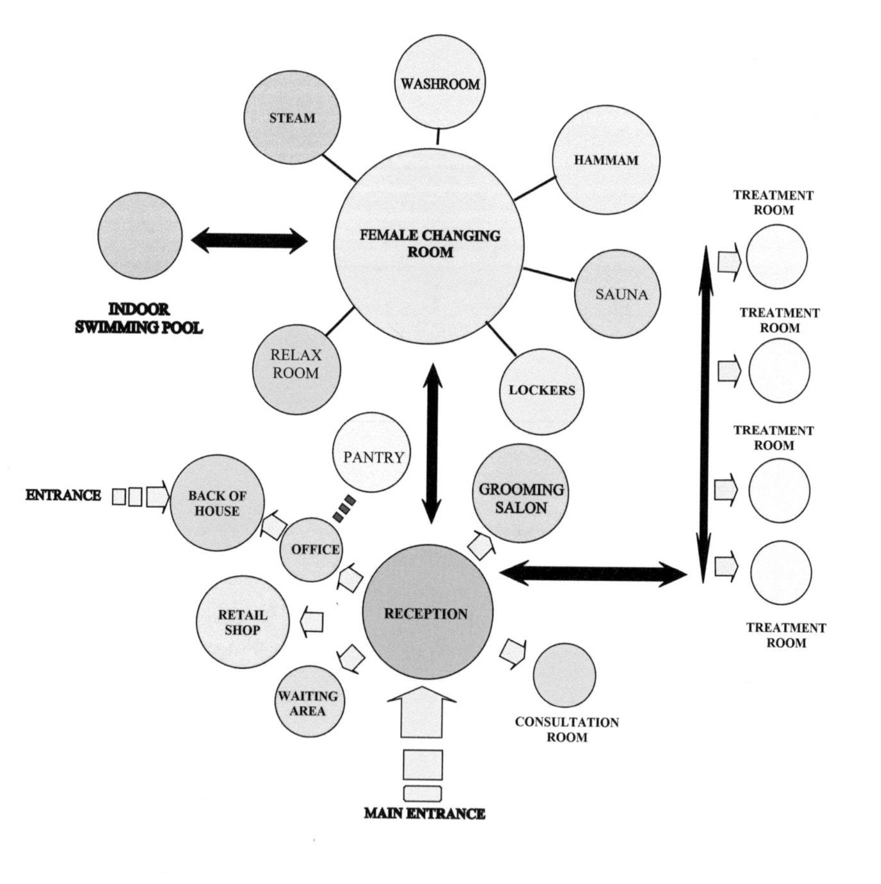

SECTION 5.0

Building Materials, Construction and Finishes

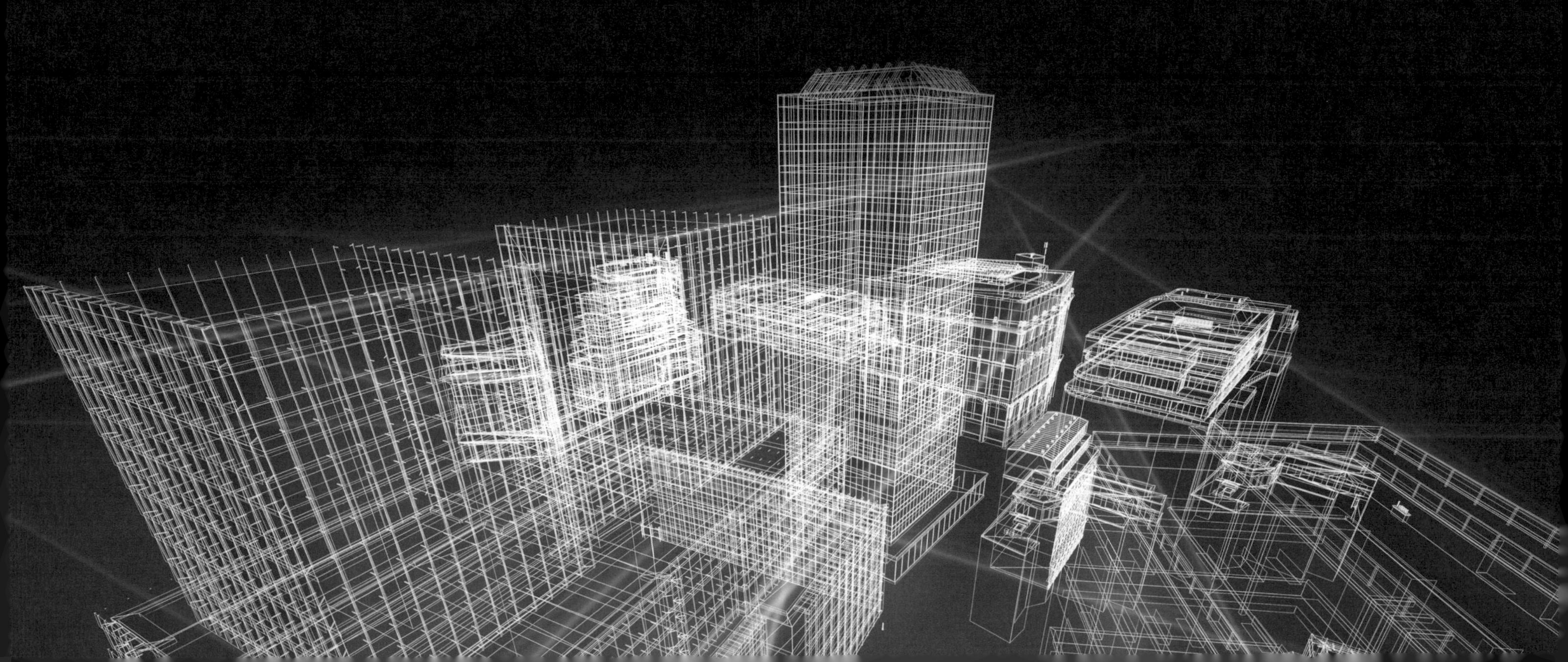

5.1.0 Building Materials and Design

5.1.1 General

There are several categories of individuals associated with the construction process who require a more rational approach to the evaluation and selection of building products. They can be identified and their area of responsibility pinpointed as follows:

Architect and Engineer: The design professional must evaluate and select building products for use in a project.

Construction Manager: The construction phase expert must make recommendations on building products to bring a project in at the construction cost estimate.

Subcontractor: The specialist in a building trade becomes a licensed applicator or exponent of specific products.

Building Materials Manufacturer: The developer of a product who must provide adequate information in the form of product literature.

Materials evaluation and selection is concerned with two major areas:

Evaluation of a new material

Evaluation and selection of materials for use in a project. New materials can be separated into two distinct categories:

- Those that are manufactured to meet an existing standard
- Those that are unique in that no standard exists against which they can be measured and therefore require an investigation and assessment as to the claims of the manufacturer.

New materials that fall into the first category are readily evaluated against a product standard. Good product standards are the result of systematic technical research efforts combined with committee work such as ASTM and ANSI standards. These standards establish suitable physical and/ or chemical properties that for the most part have a direct relation to performance. New products marketed without reference to a product standard shall be evaluated using the performance concept. This structured approach is a checklist that codifies all performance requirements by major divisions so that a workable matrix is established.

Building Materials are critical in Green Building construction, and represent a major portion of criteria used to classify or certify a green building. Understanding the building product life cycle is extremely important to understanding the procurement and use of material. In the selection of building materials, environmental impacts shall be considered in addition to the normal questions about cost, aesthetics, and durability. A primary consideration in material selection is to establish how the materials affect the building's energy performance.

The design team shall compute the baseline building performance rating according to the building performance rating method in Appendix G of ANSI/ ASHRAE Standard

90.1-2016 or USGBC approved method using a computer simulation model for the whole building project.

Additionally, the following attributes should be considered for building envelope design:

The U-Value is a measure of how much heat passes through a given material. It is generally in W/m^2k, and shows the amount of heat lost in watts (W) per square meter of material when the temperature (k) outside is at least one-degree lower. The lower the U-Value, the better the insulation provided by the material.

The R-Value measures resistance to heat transfer through conduction and expresses insulation value of material, which is the inverse of the U-Value.

The project team should specify the building envelope to achieve the lowest possible U-values as defined in this section and in any case to achieve 10% improvement in the proposed building performance compared to baseline building performance rating. Building elements forming the external walls, roofs, and floors shall have an average thermal transmittance (U-Value), which does not exceed the values shown in Table 5.1.1.

Insulation, glazing, and mechanical/electrical systems shall be selected to enhance the buildings energy performance. Additionally, the material selection process shall identify how the material will affect the health of building occupants, visitors and employees.

Other important environmental considerations for material selection shall include durability and maintenance. The amount of energy required to extract, process, transport of product shall be carefully reviewed. For example, in comparing the embodied energy office steel structure with the concrete one, the production of steel is more energy intensive than concrete production. However, steel contains high amounts of recycled steel content and the steel may be recycled indefinitely after the building is demolished.

The team of architects, interior designers, engineers and specialist consultants shall ensure that products that incorporate recycled materials are fully incorporated in the building construction process. Materials and products that are extracted and manufactured within the region, thereby supporting the regional economy and reducing the environmental impacts resulting from transportation are preferred Natural materials produced from renewable sources that do not create hazardous byproducts in their manufacture shall also be referred.

The designers shall reduce the use and depletion of finite raw materials and long cycle renewable materials by replacing them with rapidly renewable materials. Consider the materials such as wool carpets, linoleum flooring, cotton batt insulation etc.

Table 5.1.1 Average Thermal Transmittance Values

Building Component	Maximum U-Value W/m2C
External walls	0.25
Roofs	0.15
Floors	0.20
Glazed Outer Doors & Roof lights	1.30
Opaque Outer doors & hatches	0.60

Reduce or eliminate the quantity of indoor air contaminants that are odorous, potentially irritating and/ or harmful to comfort and well-being of installers and building occupants. Specify Low-VOC materials, paints and coatings, carpet products and systems in construction documents. Ensure that VOC limits are clearly stated in each section where adhesives and sealants are addressed. When choosing among various materials and products, the design team shall undertake a life cycle assessment (LCA), to systematically evaluate multiple potential environmental impacts of a product throughout its lifespan. An LCA can help identify opportunities to reduce potential impacts and minimize resource use across a product's life. It also serves to identify tradeoffs, such as whether attempting to decrease one environmental impact of a product may inadvertently result in another environmental impact.

Risk Assessment: When considering the safety and health impacts of a product, the architect shall consider how building occupants will be exposed to the product. A product such as insulation, which is installed behind a wall without direct contact with building occupants, should be evaluated differently than the vinyl flooring that occupants walk on and have direct contact with every day. Understanding how to recognize and address potential risk is critical when considering how to balance the benefits of products with their potential health, environmental and sustainability impacts.

Materials & Ingredients Lists:

The Owner will require information related to the materials and ingredients used in a building product, particularly those that come in contact with the building occupant. Additionally, product selection must meet client expectations through an evaluation of technical performance, aesthetics, cost, environmental impact, or other criteria. It is essential for all members of a design team to agree on the priority ranking so there is no misunderstanding about why a particular product is selected.

Materials and products that could be in whole or part recycled material, shall be identified:

- **Potential Recycled Material**: Masonry, Roofing Membrane, Steel, Timber, Glass etc.
- **Potential Recycled Products**: Windows, Doors, Plumbing Fittings, Electrical Accessories.

Project-specific performance of various design alternatives evaluation criteria must be defined. Since each material has many characteristics or attributes that contribute to its overall performance and its applicability to a particular project, these attributes shall be grouped by category, as shown below:

Structural serviceability: natural forces, strength properties

Fire safety: fire resistance, flame spread, smoke development, toxicity, fuel load, combustibility

Habitability: thermal properties, acoustic properties, water permeability, optical properties, hygiene, comfort, safety

Durability: resistance to wear, weathering adhesion of coatings, dimensional stability, mechanical properties, rheological properties

Practicability: transport, storage on site, handling at installation, field tolerances, connections

Compatibility: jointing materials, coatings, galvanic interaction or corrosion resistance

Maintainability: compatibility of coatings, indention and puncture (patching), chemical or graffiti attack

Environmental impact: resource consumption at production, life-cycle impact, LEED points

Cost: installed cost, maintenance cost

Aesthetics: visual impact, customizing options, color selection

5.2.0 Building Exterior and Entryway Design

5.2.1 General

Materials and Colors. Exterior building materials and finishes should convey an impression of permanence and durability. Exterior building materials and colors comprise a significant part of the visual impact of a building. Therefore, they should be aesthetically pleasing and compatible with materials and colors used in adjoining neighborhoods. Materials such as masonry, stone, stucco, wood, terra cotta, and tile are encouraged.Where masonry is used for exterior finish, decorative patterns should be considered. These patterns could include a change in color or material. Exterior colors should be given careful consideration in the context of the surrounding buildings and environment.

The materials used in design of the buildings should also reinforce the diverse experience of the elevations. Roof colors should be coordinated to complement the color schemes.

Plaza or courtyard materials shall be used to create a community space, through the use of color and scoring as patterns in the hardscape. Landscape creates diversity, provides color and softens the building and hardscape environment, while benches or seating areas, play areas and public art help residents and visitors enjoy the space and environment, making it a place where people want to visit, shop, live and recreate

5.2.2 Entryways

Entryway design elements and variations should give orientation and aesthetically pleasing character to the building. The standards identify desirable entryway design features. Entrances to residential, office or other upper story uses shall be clearly distinguishable in form and location from retail entrances, and should be:

Accented by architectural elements such as clerestory windows, sidelights and ornamental light fixtures, and/or; Indicated by a recessed entrance, vestibule or lobby, with doorways recessed for privacy and clearly expressed by awnings, high quality materials or other architectural treatments.

Each principal building on the site shall have clearly defined, highly visible customer entrances featuring a combination of the following:

- Canopies or porticos
- Overhangs
- Recesses/projections
- Arcades
- Raised corniced parapets over the door
- Peaked roof forms
- Arches
- Outdoor patios
- Display windows

Architectural details such as tile work and moldings which are integrated into the building structure and design Integral planters or wing walls that incorporate landscaped areas and/or places for sitting. In concert with the primary building material(s), a variety of materials is encouraged to articulate different building elements, such as the ground floor façade, the building base, horizontal break bands, pier or column bases, roof terminations, sills, awnings and similar building components.

Building materials should be used to differentiate between commercial and residential uses, and should create a smooth transition between the two.

The number of different materials used on the exterior of a structure should be limited to an appropriate and varied palette.

5.2.3 Materials, Construction and Finishes

Floors: Entry vestibule floors should be marble, granite, natural stone, in-situ terrazzo. Polished stone or other polished materials are not acceptable due to potential slipping hazards. In vestibules with automatic sliding or swinging doors, provide a walk-off mat in the walking area between doors. For revolving doors, provide a walk-off mat on the exterior side of the revolver. In climates where rainfall is common, a recessed mat with a floor drain is recommended.

Base: None required at walk off mats. Provide base material matching floor and wall materials at other locations.

Walls: Durable materials such as glass, stone, metal, etc.

Ceilings: Durable materials such as metal, plaster, and gypsum board should be used.

Doors: Automatic sliding or revolving doors should be provided at the Main Customer Entrance. When automatic sliding doors are provided, vestibules should be a minimum of 3.7 m deep. Swinging doors (1.07m minimum width) should be provided for exiting as required by code. Vestibules should be designed to minimize drafts in the Entrance Lobby. The revolving door would have a minimum diameter of 3.35 m. Provide safety glass in all doors and sidelights as required by code.

Revolving doors should be bronze Muntz metal, polished stainless steel or similar selection. Polished brass or bronze is not recommended due to high maintenance requirements. Swinging and sliding doors should be bronze Muntz metal (preferred), polished stainless steel or aluminum with "kynar" type paint finish.

All doors in the path of egress leading to escape stair enclosures (both internal & external and protected fire routes) shall be capable of resisting the passage of flame and smoke, and provide insulation as defined under the prescribed conditions of test appropriate to such construction, in accordance with the current version of NFPA 80. All doors on main traffic and circulation routes, corridors, lobbies or stair enclosures etc., must be provided with Vision Panels for general safety (regardless of being fire doors or not but Vision Panels should be ANSI Z97.1 certified and embedded with a fire door).

Hardware: Doors should open outward on floor hinges. Provide mortise dead locks operable by key from either side, installed in the top and bottom rail of each door. Locks should be keyed alike with all other public entrance doors, under the Grand Master key. Provide push/pull hardware on all swinging doors.

Durable, reliable mechanical door closers surface mounted, concealed in the door, overhead concealed and concealed in the floor shall fully comply with ANSI/BHMA A156.4 standard, and meet ADA Title III performance requirements. Hardware required for accessible door passage shall be mounted no higher than 122 cm above finished floor.

Lighting: Provide automated controls that either turn off or dim artificial decorative lighting (LED Luminaires) installation in the underside of the canopy to brightly illuminate the pavement area below, in response to the available daylight in the space.

***Level*:** See Engineering Design Criteria.

Power Requirements

Provide power as required for automatic door sensors and opening devices.

Special Requirements:

Provide conditioned air to prevent drafts in the entrance lobby and to eliminate condensation on entrance vestibule glazing.

Canopies over building entries shall be incorporated into the design of the building, including colors and material detailing.

Building signs should be located within an area of the façade that enhances and complements the architectural design. Signs should generally be symmetrically located within a defined architectural space, and should not obscure architectural details such as recesses, ornaments or structural bays.

5.3.0 The Main Entrances and Concourse

5.3.1 General

The main entrance to the concourse provides the customers with their first impression of the inside of the Mall. It should appear ample and unrestrictive. At the same time, it should be welcoming, comfortable and hospitable. It should also meet the operational requirements of serving as the customer's main circulation center and gathering place.

A generous first floor height shall provide prominence to the street level, establishes a clear presence for retail, and increases the visibility, marketability, and utility of ground floor space. A minimum floor height-to-floor height of 4.5 m that would accommodate the vertical clearance required for loading spaces and truck maneuvering within the rear of the building, which is essential to the viability of retail and many other commercial activities.

The design and materials of the base building shall create a warm modern environment.Wood-like materials, fabrics and indirect lighting shall filter the direct sunlight during the day and provide surfaces to light in the evening. The overall appearance should be clean and contemporary.

A rich palette of tile, carpet, warm slatted ceiling screens, fabric scrims and decorative banners shall create an upscale, comfortable shopping environment. The concourse is punctuated with soft seating areas and plantings to soften the space and provide areas for rest.

Its successful design of the main entrance sets the keynote for the entire mall and in interpreting the general tenor of the design standards which follow. Special attention should be paid to achieving the highest standards of detailing. To this end, it should be implicit that there should be full coordination from the outset between the architect, owner appointed interior designer, retail store tenant's designer and the lighting consultant, working in conjunction with the technical team. An institutional, repetitive specialty retail center look should be avoided. Customers should have barrier-free access to the concourse and retail areas.

5.3.2 Materials, Construction and Finishes

Architect should coordinate with Interior Designer and Lighting Consultant, to create a unique shopping experience through the effective choice of color schemes, layouts, displays and lighting.

Lighting:

The lighting control system shall consist of total master controls that radiate out from a central point such as a security control room. The master system shall enable at-a-glance monitoring of properties' entire lighting installations and ensure quick, efficient troubleshooting of potential problems. The lighting Control system shall meet all shopping malls' essential energy efficiency, operation and maintenance needs.

The lighting system shall integrate all control features associated with lighting via a centralized intelligent control panel or a distributed system. Specific inclusions in the systems shall include occupancy detection, dimming, daylight linking and scene-setting. All controlled areas and loads shall be displayed on a PC across multiple windows via labels initiated by the Mall operator with current load statuses shown in real time. Programmable Dimmers shall be provided to enable architectural lighting goals, in addition to creating extraordinary lighting effects through the control of large numbers or down and display lights. Scene Controls shall be used to operate several device series and set multiple levels or actions via one command and typically used to create mood lighting and quickly transform in-mall ambience. Schedule-switch or dim several lighting areas at a day's start or end. Enable/ disable occupancy sensors when moving from working to nonworking hours' control.

The specialist lighting designer/architect shall evaluate alternative placement and massing concepts for the individual building sites at the scale of the block to secure the greatest amount of sunlight and sky view in the surrounding context. All public area lighting requirements shall be capable of being activated via a combination of manual switches and/or automatic controls via time scheduling or light level. Light level sensors shall determine lighting levels according to the sunlight filtering through windows in specific areas. Should a dimmer detect less than optimal artificial brightness in an area, it shall brighten lights to the desired brightness value. Similarly, when natural brightness begins to fade outdoors, the lighting control system shall adjust the artificial light in small steps. Should the Mall operator wish to, users shall be able to override these automatic light controls by means of a simple push-button. Lighting should be switched from panels located within an employee area not accessible by customers.

Level: See Engineering Design Criteria.

Power Requirements

Power should be provided as required by code. Outlets should be provided for housekeeping equipment (13A 240V) throughout public and circulation areas. They should be placed not greater than 15 m o.c. All receptacles should be mounted 40 mm A.F.F. or coordinated with base materials. Outlets for specific uses should be as located by Interior Design requirements.

Signage and Wayfinding

The audio-visual design team shall provide a software driven solution offering all possible functions of an efficient, flexible wayfinding guidance system to guide guests to specific shops, departments, and floors while improving experiences by providing them instant access to on-demand information.

The system should be capable to seamlessly blend the malls branding, facility wayfinding, tenant sales, food court menu, into a compelling, easy to manage, all-in- one visual communications platform. In combination with reading devices (barcode, swipe card, chip and so on), functions may be expanded.

The system of wayfinding shall be through the use of digital signage, informational displays and kiosks. Wayfinding shall also include audible communication, tactile elements, and provisions for other persons with disabilities.

The system shall have the following features:

- Built-In keyboard for quick and easy destination searching.
- Directions automatically generated once a destination is selected.

- 'You Here' map icon with turn by turn directional arrows.
- All directions shall be based on shortest path & ease of accessibility (Smart Pathway Algorithm)
- Directions can be chosen to be sent to a mobile device via text message or email.

Implementing an effective wayfinding system will ensure that customers have a positive experience and do not need staff assistance to direct them to their destinations. Signage shall be consistent in appearance and application, allowing the viewer to become familiar with the system and use it efficiently.

General guidelines for sign placement:

- Locate signs at or near decision points with consideration for the viewer's path of travel
- Place signs perpendicular to path of travel
- Place signs for maximum visibility within comfortable field of vision and viewing angle
- Place signs at consistent heights and on uniform mounting surfaces whenever possible
- Locate signs with sufficient frequency to ensure viewers do not feel abandoned

Sound System: See Audio-Visual Systems

5.4.0 Retail Stores

5.4.1 General

All retail stores must be designed by the Tenant's Architect in accordance with the Design Criteria provided by the Owner and shall be subject to their review and approval process. Compatibility with adjacent and previously approved store designs shall be considered as part of the overall review process. The first 183 cm behind the storefront is considered the Tenant's Merchandising Zone and will be subject to strict Mall operator's design controls. To allow each store design maximum visual impact, storefronts shall be framed on both sides by owner provided 25 cm wide demising neutral piers.

Tenants shall be encouraged to use materials and methods in the design of their storefront that enhance the overall design intent of the Mall. A wide variety of materials are acceptable; however, they must be executed in a clean and contemporary style. Tenants shall creatively engage the customer by incorporating digital media into the storefront and interior design.

The store façade, interior design and general construction shall reinforce the mall's design ideals, and be comparable in quality with the overall shopping center.

Depending on their distance from the public washrooms, Tenants may be required by code to provide washroom facilities within their spaces. It shall be the responsibility of the Tenant's architect to ascertain the specifics of this requirement. All Tenant washroom facilities must be ADA/ DDA compliant.

Figure 5.4.1 Retail Store

5.4.2 Materials, Construction and Finishes Storefront:

Storefronts shall be predominantly glass, and emphasize well-designed architectural elements, a sense of entry and imaginative show window displays.

Glass – Clear tempered with polished and beveled edges, sandblasted or etched glass.

Solid core materials - Vitrastone, PaperStone, etc. or equal. Metals – stainless steel, anodized aluminum, bronze, copper, etc.

Tile – mosaic, natural stone, terrazzo.

Wood – finish grade hardwoods, painted or stained

Plaster – smooth finish

Additional materials are encouraged – brick, cast stone, steel elements.

Floors: The tenant shall match mall border tile from the lease line to the Tenant's storefront line and to all recesses up to the line of closure. Provide a set minimum of hard surface transition flooring material past the lease line. Provide hard surface flooring materials (stone, ceramic pavers, etc.) with a minimum coefficient of friction of 0.55 - 0.6 per ASTM C 1028, accented with area rugs. Carpeting should be contract grade 10 row - Axminster 80/20 wool/ nylon, and a cotton poly backing. A polypropylene backing should be required in high humidity areas. All carpeting should be laid on a carpet pad, compliant with the Carpet and Rug Institute (CRI) Indoor Air Quality Cushion Testing Green Label Program. Changes in elevation should be indicated by different floor finishes or changes in colour. Special attention is required at transitions between hard surface materials and carpets to protect the edges of hard surface materials and to prevent tripping hazards. Provide metal strips or butt jointing between all floor material changes.

In all areas with water, such as kitchens, washrooms and water features, Tenants shall install a waterproof membrane under their floor finish and 15 cm up the demising walls. Flood testing of this membrane will be required for the Owner's review and approval.

Base: Base (150 mm) is required on all walls; base material should be appropriate for the selected floor material as determined by the tenant's Interior Designer.

Walls: Wall materials should be selected by the tenant's Interior Designer. "Architectural-type" finishes (brick, marble, fabric wrapped panels, wood, etc.) are preferred, appropriate to the design and volume of space. Fabric backed vinyl wallcovering, where provided, should be Type II, minimum 20 oz. weight. Wall mounted shelf standards must be fully recessed into the adjacent gypsum board and may not compromise the fire rating of a demising wall.

All demising walls are to be furred out. Nothing can be attached to the demising wall. Fixtures must all be attached to the furred-out walls. All wall construction and column enclosures must conform to the applicable local fire and building codes. All demising walls shall be of a construction that maintain a one-hour fire rating. Gypsum board shall extend from the floor to the deck above.

All music, video, hair & nail salons require a minimum 5 cm sound insulation.

Ceiling: All ceilings in the Merchandising Zone shall be gypsum board extending a minimum of 125 cm back from the Tenant's storefront glass. Ceiling heights in the tenant's space shall be a minimum of 305cm.Acceptable materials include gypsum board, plaster and metal panels.

In gypsum board ceilings, Tenants are required to provide 46 cm x 46 cm hidden flange metal panels flush at the perimeter to allow access to Mall operator's and Tenant's mechanical systems.

Locate exit sign behind ceiling soffit so as not to be seen from the mall's common area. Mechanical supply and return air grilles should be linear diffusers, decorative grilles or hidden from view.

Sprinkler heads shall be flush or fully recessed, and the cover plates are to be chrome or to match ceiling color. Locations should be coordinated with lights, speakers, etc. Speakers must be located a minimum of 610 cm from storefront lease line

Doors: Regardless of the direction of door swing, the door shall be recessed a minimum of 46 cm behind the lease line. Open doors shall be fully recessed behind the lease line.

As an alternate, sliding glass door may be used. Where utilized, door framing, mullions and jambs shall align when the doors are opened, and the storefront shall be designed to minimize the appearance of such doors.

Storefront doors shall be equipped with emergency quick release locks as required by Code. All service corridor exits must be one hour rated steel doors and frames with automatic closers that fully comply with the latest version of NFPA 80.

Lighting/Sound:

Lighting systems should be highly flexible, enabling a fresh, up-to-date appearance, and accommodate products, offers, and displays that change frequently. Provide a system with one-button activation of lighting, music and audio-visual displays that operate a mix of localized light, music and master controls.

Provide a single channel Audio Distribution Unit (ADU) which will enable the amplifier to retain both a local MP3 player and a secondary input, and have the ability to use Infra-red (IR) reticulation via the amplifier. The system shall provide on/off and volume controls via or touchscreen. Where multi-room configuration exists, multi-source music shall be tailored to provide different music to specific areas at varying volumes.

Recessed lighting fixtures using reflectors, baffles, and lensed trims shall be used for general lighting. Down lights shall be carefully designed to draw attention to labels and price tags. All adjustable units shall be positioned so that the light is not directed into the Mall Common Area. Fixture quantity and locations are subject to Mall operator's review and approval.

Storefronts which are recessed behind the lease line shall have tenant's soffit area, both inside and outside the closure line, in the form of recessed downlights. Fixtures shall provide baffled glare-free light. Fixture quantity and locations are subject to Mall operator's review and approval.

All signage and show window illumination shall be controlled by a time clock and illuminated during Mall operating hours. The tenant shall provide lighting for the entire leased premises. All store cases shall be adequately illuminated and vented.

Energy efficient Fluorescent fixtures in areas visible to the public shall be recessed and shall have deep cell parabolic reflector lenses. Fluorescent lamp color shall be "Deluxe Warm" or other colors subject to Mall operator's review and approval. Lamp color shall be consistent among like fixtures throughout the sales area.

All store lighting to be glare free.

Signage:

Each Tenant shall design, fabricate, install and maintain one primary sign on their storefront facade. All signage should be visually consistent with the store interior design. All signs must be submitted to the Owner for review and approval in the form of shop drawings and photographs or samples if applicable. All shop drawings must show the sign placed on the full storefront elevation. Shop drawings must contain all applicable information about the sign including letter size, depth, materials, colors and method of support. All signs must be "UL" approved.

Acceptable Sign Types: Individual channel letters, Backlit individual letters, Routed wood, metal or stone, Metal letters mounted in front of or behind glass, Wood letters covered with gold leaf or silver leaf, Sandblasted glass, Dimensional letters of metal, stone, or painted wood.

Signage shall be limited to Tenant trade name only as defined in the lease. Signage shall not exceed 70% of the width of the storefront.

Tenant must provide concealed local disconnect switch at each sign location.

One primary identity sign shall be allowed per storefront. Any Tenant with more than one exposure may incorporate one sign per storefront elevation at the discretion of the Mall operator. Exterior frontage tenants could have an additional secondary identity sign just above entry. Interior signage letters shall not exceed 40 cm in height and 10 cm in depth. Exterior signage letters shall not exceed 61 cm in height and 15 cm in depth. Any additional store identification signs within the tenant space must be no closer than 122 cm from the storefront window.

All illuminated signs must be fabricated and installed according to national, local building and electrical codes and must bear UL label.

5.5.0 Full-Service Food Court Criteria

5.5.1 General

The Food Court tenants will be located around a common eating area where tables and seating shall be provided. The Architect shall design the area to have Transom windows above that will allow natural light to create an open, attractive atmosphere.

Tenants shall be encouraged to develop a unique and innovative design and not a traditional franchise look. It is intent of the owner that the Food Court tenants convey an upscale food market atmosphere, and therefore it is important that the Tenant's design shall be compatible with neighboring Tenants.

All Food Court tenants must comply with all health regulations and applicable ordinances of local governing authorities.

The Design Criteria provide a framework with sufficient flexibility so that national or regionally recognized food retailers can adapt their design concepts to fit the requirements of the Owner The Design Control Zone shall extend from the Lease Line to and including the face of the wall separating the sales area from the food preparation area within Tenant's demised premises.

The Owner shall approve all Food Court Tenant designs, signage and materials within this zone. Emphasis in this area should be on imaginative use of materials, attractive food display and sales.

No means of security closure shall be permitted between the Design Control Zone and the Lease Line. Tenants will be required to break down their Tenant front operation, clean, and store any removable equipment in lockable cabinets or in a secured rear area.

Each Tenant shall be required to engage the services of an architect. Store plans must be drawn to comply with all applicable National and International Building Codes, as well as all applicable ADA/DDA guidelines.

Depending on their distance from the public washrooms, Tenants may be required by the local Health Department to provide separate employee toilet facilities within the premises. These facilities must be ADA/DDA compliant.

In the "back of house" areas, all finishes must meet all applicable National and International Building and Health code requirements. All ceilings must be non-combustible and meet fire code requirements.

5.5.2 Materials, Construction and Finishes

Counters: Tenants must provide a continuous front counter across the full width of each storefront. Access doors are permitted only where there is no back entry provided. Otherwise, all service must be through the Tenant's back door.

Tenants shall design their front counters to contain areas for food display. Counter must comply with ADA/DDA requirements. Sneeze guards, where required shall be architecturally compatible with counter design and in accordance with Code. All back counters facing the public must have a finished look with lower cabinets and doors. Acceptable Food Court counter top materials include:

- Solid surface plastic materials such a "Corian" at least 1/2" thick.
- Granite or granite composites
- Brushed finish stainless steel
- Ceramic tile
- Brass

Acceptable Food Court counter face materials include:

- Finished solid hardwoods
- Brushed finish stainless steel
- Porcelain panels
- Granite or marble
- Ceramic tiles
- Glass blocks or tiles

(a) **Floors:**

The Owner will provide a finished floor in the Food Court common area. The Tenant shall be required to extend the Landlord floor into any recesses created by their front counter design. The Tenant's flooring within their space should be a durable ceramic, quarry or stone tile. All floors must have a waterproof membrane under the tile within the demised space. This membrane must extend 30cm up all demising walls.

(b) **Walls:**

The decorative side demising walls within the Merchandising Zone are provided by the owner, and may not be altered. The front Merchandising Zone shall be separated from the kitchen/back of house area with a full height partition wall. Open kitchens are not permitted. All wall construction must conform to the applicable fire codes.

All wall surfaces must be washable as per local Health Department requirements. All demising partitions must be one hour rated metal stud construction with sealed penetrations. The walls shall extend from the floor to the deck above.

(c) **Ceilings:**

Ceilings in the Merchandise Zone shall be gypsum board painted to match the base building color. Tenant ceilings within the closed kitchen areas may be drop in panels with a vinyl finish to meet local codes and Health Department requirements. The drop panels should have a non-perforated cleanable surface with a non-reveal edge. Drop in panel ceilings or acoustical ceiling tiles are not acceptable in the Control Zone Area. All ceilings must be non-combustible and meet all fire code requirements.

(d) **Doors:**

Tenants shall use full height metal doors going into their kitchen areas. Partial height doors or swinging gates will not be permitted. All doors leading into the back of house must be equipped with automatic closers.

All doors to the service corridor must be steel with fire rated doors and frames.

Doors from the Tenant "back of house" into the service corridor must be recessed so they do not swing outward beyond the lease line and block the required clear exit width of that corridor.

(e) **Lighting:**

The Tenant shall be required to illuminate adequately the countertop with recessed energy efficient fixtures. The use of hidden cove lighting, low voltage accent lighting, recessed LED fixtures, down lights and decorative pendant lights is encouraged. Fixtures visible to the public are subject to Owner approval.

Fixtures used for lighting the menu board and counter tops shall be controlled by dimmers. The use of fluorescent lighting shall not be permitted.

Emergency Lighting: Each Tenant shall provide emergency lighting in accordance with all applicable codes to emergency exits. All such lights should be clearly marked on Tenant's electrical panels.

Level: See Engineering Design Criteria.

5.6.0 Public Lift Lobbies

Passenger Lifts shall be grouped in banks of at least two for efficiency. Lift groups of four or more shall be separated into two banks opposite each other for maximum efficiency in passenger loading and minimum hall call notification for accessibility under requirements of ADA/DDA.

Travel distances from a given office or workstation to a lift should not exceed 61 m. The positioning of vertical circulation elements (escalators and lifts) require great care to avoid "dead-ends" and "double-back" circulation. These elements should be provided in the natural circulation paths of office staff and shoppers.

In general shopping malls are designed to occupy two levels. Two levels are generally considered as much as the average shopper is prepared to contemplate. Based on the Merchandising Plan, a third level may be added. Malls with three levels shall have food courts at the upper or lower levels to form an attraction and a contrast to the main sales areas.

In shopping malls lifts are used to transport shoppers to/from car parks and to allow shoppers with prams and pushchairs to access all levels. Observation lifts may be used for the latter purpose.

5.6.1 Program Requirements

Facilities Included:

- Main Level Public Lift Lobby.
- Secondary Level Public Area Lobbies.
- Office/Residential Apartment Level Lift Lobbies.

Facilities Program:

The planning of lifts and their related facilities shall take into consideration the special requirements of the public areas in relation to building circulation, both horizontal and vertical. The specialist lift designer and the architect shall consider various aesthetic and conceptual ideas early, and establish the optimal solution for building.

The net effect should be a building properly configured for good access with sufficient handling capacity to serve the proposed population and its circulation needs. Where the building has more than one main entrance, each entrance may be served by its own group of lifts.

The interior design finishes of the public lift lobby should complement the design of the main entrance and concourse with wall materials preferably of stronger, warmer colors to attract attention to this most important center point of public circulation. The ceiling should be lower than the general lobby ceiling level, with special lighting treatment of somewhat higher intensity to further accentuate its important location.

Minimum Requirements: Provide a minimum of 3.0 m between opposing banks of Lifts, and 2.5 m for a lobby serving a single bank of lifts.

5.6.2 Materials, Construction and Finishes

(a) **Floors:**

Provide hard surface flooring materials (stone, ceramic pavers, etc.) with a minimum coefficient of friction of 0.55 - 0.6 per ASTM C 1028.0R DIN 51097. Attention is required at transitions between hard surface materials and carpets to protect the edges of hard surface materials and to prevent tripping hazards.

(b) **Base:**

A 150-mm base is required on all walls in public areas; base material should be appropriate for the selected floor material as determined by the Interior Designer.

(c) **Walls:**

Provide "architectural-type" finishes (brick, marble, fabric wrapped panels, wood, etc.).

(d) **Ceiling:**

Ceilings should be designed as an extension of the concourse ceiling system with provisions for soffits, light coves, etc.

(e) **Doors:**

Lift doors and frames should be finished metal that would complement the interior design concept.

(f) **Lighting:**

The primary light source should be solid state lighting with dimming.

Level: See Engineering Design Criteria

(g) **Power Requirements:**

Outlets should be provided for housekeeping equipment (13A 240V) throughout public and circulation areas. They should be placed not greater than 15 m o.c. All receptacles should be mounted 40 mm A.F.F. or coordinated with base materials. Outlets for specific uses should be as located by Interior Design requirements.

5.6.3 Public Lift Cabs

Program Requirements

Facilities Include: Lift cabs for public use.

Facilities Program: Lift cabs should be designed as an extension of the main concourse in terms of quality and finishes.

Minimum Requirements: Lift cab design should meet ADA/ DDA and all local codes.

5.6.4 Materials, Construction and Finishes

(a) **Floors:**

Provide hard surface flooring materials (marble, pavers, etc.). Flooring should be an extension of the main concourse floor materials.

(b) **Base:**

A 89 mm high base shall be fitted between the finished floor and the wall panels. Bases shall consist of Stainless Steel or Fused Metal decorative faces applied to an extruded aluminum substrate.

(c) **Walls:**

Wall panels shall consist of graffiti resistant extruded aluminum frames surrounding insets with a decorative face material on a fire-rated substrate. Handrails should be provided on at least one wall as required by code.

(d) **Ceiling:**

A multi-panel suspended island design or other ceiling concept with low-voltage, high-efficiency LED downlights or high-output LED perimeter lighting. The level of illumination at the car control panel shall be a minimum of 54 lux. Ceiling panels shall consist of Stainless Steel or Fused Metal decorative faces applied to a structural backer with black reveals.

(e) **Doors:**

Lift doors shall be in matching materials and finishes to lift interiors.

(f) **Special Requirements:**

A speaker should be provided for background music and emergency messaging. Two operating panels are required, one on each side of the Lift door. Advertising signage, if provided, should comply with mall operational standards.

Provide an emergency communication system in each cab, which shall fully comply with ADA/DDA requirements. Provide floor numbers on the Lift panel in Braille and raised numerals as required by ADA/DDA.

5.7.0 Public Washrooms

The design team shall take elements from the rest of the building's interior design or the surrounding environment and bring it into the washroom for a more coherent and seamless look. The washroom space built with superior materials shall have a relaxing, almost therapeutic design aesthetic.

To enhance security particularly for females, entrances to washrooms should be located where the clearest site- line to high traffic public areas exist. Labyrinth entrances directly along major traffic corridors provide both the 'sense of' and actual security. A configuration that locates washrooms at the end of long corridor's, where users have no acoustic or visual site-line to trusted persons is not acceptable.

Provide background music inside the washroom, to besides improving the ambience, shall provide a level of acoustic privacy.

5.7.1 Facilities Include Public Washrooms for Handicapped Visitors

Washroom Location

Public Washrooms of sufficient capacity is to be located conveniently to all public circulation areas, food court, meeting facilities etc. They should, wherever possible, be located at the same level as the spaces they serve, and in no case, should the level of such washrooms be more than 3 m above or below the level of such spaces.

Wherever possible, locate Washrooms to serve more than one group of facilities. They should, however, be accessible to visitors using the food court, be close to retail outlets and be easily detected by the customers.

Handicapped Visitors: Provide a special washroom for handicapped visitors at each level on which retail outlets are located. Such washrooms should open off a public corridor to permit easy use by individuals rather than be contained within the general toilet areas.

Washroom facilities for entertainment lounges, fitness centers and/or swimming pools should be separate from all other uses. The number and type of sanitary fixtures should comply with local regulations in all respects.

Planning: Public Washrooms should have vestibules arranged to screen the interior spaces from public view. Female washrooms should be provided with a make- up room between the entrance and the toilet area, with adequate lighting suitable for making-up.

Arrangement of fixtures should be such that visitors pass the washbasins naturally when leaving the WC/urinal areas. WC enclosures should be 900 mm wide x 1500 mm long, as a minimum. In each main group of Public Washrooms, provide at least one cleaner's closet, not opening off a toilet area itself, finished and equipped in accordance with the standards established for Janitors Closets.

5.7.2 Materials, Construction and Finishes

(a) **Finishes:**

Public Washrooms should maintain the same high level of design and finishes as are provided in the areas they serve, and the Architect, Interior Designer and Lighting Consultant should cooperate fully towards this end.

All washrooms should have moisture impervious cleanable surfaces. The finishes must be sufficiently durable to withstand the anticipated traffic levels and the washroom- cleaning frequency.

(b) **Floors:**

Marble, natural stone, unglazed ceramic tile, or homogeneous tiles, sloped to floor drains in male and female washroom fixture areas, with a minimum coefficient of friction of 0.55 - 0.6 per ASTM C 1028.0R DIN 51097.

(c) **Base:**

A base is required on all walls. Base material should be appropriate for the selected floor material (i.e., marble at marble flooring) and as determined by the Interior Designer.

(d) **Walls:**

Marble, natural stone, homogeneous tiles, stainless steel, glass block, phenolic cladding. or glazed ceramic tile with coved base in male washroom and entrance vestibule and in female washroom fixture areas. Walls within 610 mm of urinals and water closets should have a smooth, hard, nonabsorbent surface to a height of 1219 mm above the floor, and except for structural elements, the type of materials used in such walls should be resistant to moisture.

Public washroom walls should provide a minimum S.T.C. rating of 51 to all adjoining spaces.

(e) **Ceiling:**

Ceilings should be suspended mineral fiber board, fibrous plaster board, gypsum board with semi-gloss painted finish.

(f) **Doors:**

Entry doors should be 926 mm x 2135 mm x 44 mm solid core wood doors, with finishes appropriate to location. Frames should be integral steel or wood as appropriate. Doors should be equipped with concealed closers, push plates, pull plates and kick plates in approved materials and finishes. Outer doors of entrance vestibules should be provided with standard insignia for washrooms, and should be equipped with dead locks keyed under a Master Key, to lock up toilet when under maintenance.

The main entrance to the washroom shall be designed such that cubicles, urinals and mirrors are away from the line of sight from the main entrance.

(g) **Hardware:**

Provide push/pull hardware on entry doors.

(h) **Mirrors:**

Provide backlit large mirrors across full width of vanity shelves, and over female washroom make-up counter, extending from height of 900 mm above floor to a minimum height of 1850 mm; or a special design incorporating mirrors and illumination as a decorative treatment (especially in make-up area). Also, provide a full-length mirror in each Public Washroom, or vestibule located so as to provide each departing visitor with a view of his or her complete appearance.

(i) **Lighting Type:**

All public washrooms should be provided with warm- color lighting for general lighting as well as down lights above the wash basin/mirror. The primary light source should be LED energy saving long life bulbs. Lighting should be connected to an un-switched circuit.

Level: See Engineering Design Criteria.

(j) **Power Requirements:**

Provide convenience outlets, one per washroom (13A 240V). Locate convenience outlets below the vanity, out of public view, or in the vestibule.

(k) **Special Requirements:**

Floor Drain: Provide one waterless odor trap type floor drain, located under the toilet partitions, in each washroom.

Vanities: Vanity tops should be solid surface material or natural stone with cut-outs for under counter mounted lavatory bowls and 100 mm splashes on three sides. Provide an apron of the same or complimentary material to the countertop of sufficient depth to hide the plumbing below or another recessed screen. Recessed screens should be demountable for plumbing access and maintenance. Insulate or enclose all piping below vanity at accessible Washrooms.

Toilet Partitions: Toilet partitions should be ceiling mounted solid phenolic or plastic laminate with matching doors. WC cubicles shall be a minimum 900 mm (width) x 1500 mm (length). Cubicles should be provided with easily closable free-swinging doors. Doors should be minimum 610 mm wide, and fitted with latches, sliding dead-bolts or other similar locking devices. While door locks should be accessible from the inside only, authorized outside key access shall be necessary in emergencies or to take an out-of-order cubicle offline. All cubicle partitions should extend from floor to ceiling.

Urinal Screens: Urinals should be separated by modesty boards of not less than 300mm x 1800 mm (Height) to act as a visual barrier between urinals. These screens should be wall mounted and match toilet partitions.

Ventilation: Provide a ventilation system that ensures vitiated air is quickly extracted, and helps to avoid dampness and subsequent growth of mold on floors and walls. It is most important that proper provision be made for the transfer of sufficient air from adjacent public spaces that are permanently air-conditioned, to permit the toilet exhaust to extract the designed amount of air. The replacement air may be drawn through louvres in the doors, cuttings under the door, or other openings.

The design engineer shall ensure to provide cross ventilation and the air exchange rate should have a minimum of 15 air changes per hour. Air locks should be incorporated to separate the toilet areas from food consumption or preparation areas.

Accessories: All accessories should be stainless steel bright finish units of approved pattern as follows: -

- Double toilet paper holders in each WC enclosure
- Automatic liquid soap dispenser beside each washbasin. Locate dispenser to drain into washbasin.
- Stainless steel finish Electronic Hand-Dryers/Paper Towel dispenser and disposal unit, recessed type.
- Sanitary napkin dispenser in female washrooms, recessed or surface-mounted, next to each towel unit and individual disposal unit in each WC enclosure.
- Diaper Changing Station.

Plumbing Fixtures:

Washbasin: Under counter mounted vitreous white china bowl.

Trim: Polished chrome (minimum; chrome and brass preferred) thermostatically temperature controlled faucet with 203 mm spread, sensor control, screen drain and flow restrictor.

Water Closets: Vitreous china, one-piece elongated siphon jet bowl with white open front plastic seats and check hinges. Provide a hand-held bidet sprayer complete with squeeze trigger, T-adapter and shut-off valve, located within the convenient reach of users.

Urinals: Urinals should be actuated by individual self-activated proximity control units.

5.8.0 Prayer Rooms

The number and location of quality prayer facilities will depend on the size and architectural design of the building. The greater the distance required for customers to reach the prayer facility, the better the argument for having more than one facility. Linear and a long plan building tend to justify a non-centralized location, whereas radial plan building justifies a central location.

Males and females pray separately, and it is therefore essential to have separate, but adjacent, facilities for each gender.

5.8.1 Facilities Include:

- Female Prayer Room
- Male Prayer Room
- Ablution Area

5.8.2 Materials, Construction and Finishes

(a) **Floors:**

Provide a carpeted floor with required number of prayer mats. Provide an area between the prayer room and ablution area with hard flooring such as marble or unglazed ceramic tile with a minimum coefficient of friction of 0.7 – 0.84 per ASTM C 1028.0R DIN 51097.

(b) **Base:**

Base material should be appropriate for Carpet/hard flooring.

(c) **Walls:**

Painted in neutral color.

(d) **Ceiling:**

Suspended gypsum board with semi-gloss painted finish. An "AlQibla" direction shall be installed. This feature should be directed towards Makkah.

(e) **Lighting Type:**

The primary light source should be LED energy saving long life bulbs. Lighting should be connected to an un- switched circuit.

Level: See Engineering Design Criteria.

(f) **Power Requirements:**

Provide convenience outlets, two per room (13A 240V).

(g) **Special Requirements:**

Provide an Armoire or a closet that is equipped with a shelf to place the Quran.

5.8.3 Ablution Area

In line with religious rulings, ablutions need to be performed before performing prayer. The ablution area shall exist inside the clean zone, and the ablution space shall be located adjacent to the prayer room.

Facilities Include:

- Female Ablution Room
- Male Ablution Room

Facilities Program: A male and female ablution room should be provided, and directly connected to the respective prayer room.

5.8.4 Materials, Construction and Finishes

(a) **Floors:**

Hard flooring such as marble or unglazed ceramic tile with a minimum coefficient of friction of 0.7 – 0.84 per ASTM C 1028.0R DIN 51097. Floor should be sloped towards a suitably sized trough type floor drain

(b) **Base:**

Base material should be appropriate for the flooring.

(c) **Walls:**

Marble or glazed ceramic tile with coved base to a height of 1500 mm, and painted in neutral color to ceiling level.

(d) **Ceiling:**

Suspended gypsum board with semi-gloss painted finish.

(e) **Lighting Type:**

The primary light source should be LED energy saving long life bulbs. Lighting should be connected to an un- switched circuit.

Level: See Engineering Design Criteria.

(f) **Power Requirements:**

Provide convenience outlets, two per room (20A 240V)

(g) **Special Requirements:**

- Provide an adequately designed ledge to store footwear
- Provide a communal trough with suitably designed drainage, and a conveniently located faucet for water supply.
- Provide a properly located and constructed sitting ledge directly in front of the communal trough to facilitate washing.
- Provide storage facility for clean towels and soiled towels

5.9.0 Security Control Room

Special Requirements:

The provision of an appropriate working environment is essential for control rooms that must accommodate auditory and visual tasks. Where speech communication is important, all auditory needs of the environment must be appropriately specified. Ambient noise levels need to be controlled by considering such factors as room and console finishes, noise output of equipment, and control of external sources of noise.

The air-conditioning system shall be designed to automatically set to increase the ambient temperature to compensate for natural early morning drops in body temperature. Room temperature should range from 21°C - 23°C, with relative humidity from 40 - 65 percent and minimal air movement not exceeding 10 – 15 cm per second.

(a) **Floors:**

Provide resilient flooring material such as vinyl tiles or modular carpet tiles. Carpet Tiles shall be tested for meeting the Electrostatic Propensity requirements. Floor materials should have a lower reflectance of 0.2 to 0.3 for carpet or 0.25 to 0.45 for floor tiles.

(b) **Walls:**

Walls should be an off-white matte or flat finish with a reflectance range of 0.5 to 0.6.

(c) **Ceiling:**

Ceiling materials should offer moderate to high reflectance of 0.8 or more to improve light distribution throughout the room. Ceiling acoustics should be selected to achieve an NRC of 0.65 to 0.75 or better and an AC (articulation class, a measurement of sound attenuation) of 40 to 44 or better.

(d) **Doors:**

An approved entrance door and frame unit fitted with associated security devices and equipment shall be provided. The door and frame assembly shall comply with ASTM E () and ASTM E413.

(e) **Lighting:**

The lighting scheme shall be based on projecting ambient lighting up, creating general illumination without high levels of contrast, resulting in a soft, diffused light with almost no glare. Task lighting shall be integrated into custom consoles and aimed onto work surface and documents. Lighting should be such that it does not create veiling reflections on operator consoles or other reflective surfaces that require monitoring.

There should be no perceptible flicker from LED lighting. In designing a lighting scheme, attention needs to be given to the range of tasks undertaken as well as the ages of the operators. It is essential to provide adjustable lighting for control rooms that are manned 24 hours a day. During night-time operation lighting, shall be dimmed.

Table 5.1.4 IESNA Illuminances for Retail Lighting Design

Area/Tasks	Description	Area Type of Activity	Lux Illuminance
Circulation	Area not used for display or appraisal of merchandise or sales transactions	High Activity	325
		Medium Activity	215
		Low Activity	110
Merchandise (Including Showcases and wall displays)	That plane area, horizontal to vertical, where mechandise is displayed and readily accessible for customer examination	High Activity	1075
		Medium Activity	805
		Low Activity	325
Feature Displays	Single item or items requiring special highlighting to visually attract and set apart from the surround	High Activity	5380
		Medium Activity	3230
		Low Activity	1615
Show Windows - Daytime Lighting			
General			2150
Feature			10,765
Nighttime Lighting			
General			1075 - 2150
Feature			5380 – 10,765

SECTION 6.0

Recreational Facilities

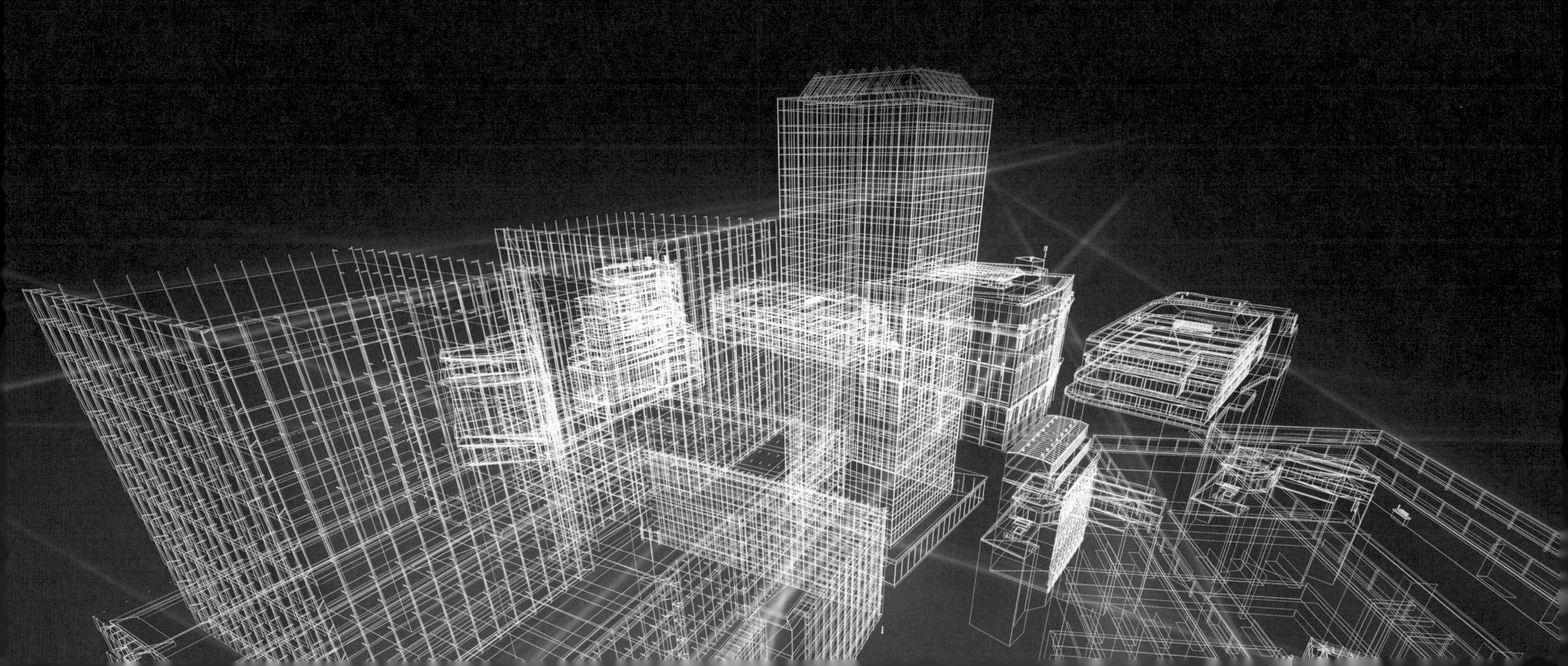

6.1.0 Swimming Pools

6.1.1 General

These standards are intended for application to public pools which are used strictly for recreational purposes by registered members.

Design features that are not specifically covered by these standards shall be permitted only where it is demonstrated that adequate safety and water quality can be maintained, based on current technology and the best information available at the time.

Facilities Include:

- Outdoor Pool
- Indoor Swimming Pool
- Pool Snack Bar
- Pool Locker Rooms

Facilities Program:

Outdoor pools for exercise and/or recreational swimming as appropriate for the market location should be positioned to receive direct sunlight from 11:00 a.m. until 5:30 or 6:00 p.m. throughout the season when the climate permits their use (except that in areas with extremely hot climates, where the direct rays of the sun are unbearable, the program may specify that the pool be partially shaded, and the water cooled).Local customs require that the swimming pool not be in view of the guest areas in general. In such cases, the program should require that the pool be screened from any view outside the immediate pool area.

Access to the outdoor swimming pool and pool terrace area should be so arranged that the registered members enter the area through a single passage which can be controlled by the Male Fitness Center reception. The separation between this area and the remaining grounds and terraces need not be more than symbolic, such as a low wall or hedge and a change of level, to indicate where free passage is not desired. Access to the indoor swimming pool should be directly from the Female Fitness Center reception.

Minimum Requirements: The pool should be of an interesting, irregular shape, and of a minimum surface area equal to a rectangular pool 25 m X 15 m in size. The shape should be such as to provide two parallel sides approximately 10 m long and 20 m to 25 m apart, which can be used for swimming back and forth for exercise. The maximum depth should be 2.5 m and the minimum depth 1.1 m. The bottom of the pool should slope toward the main drain. Where the water depth is less than 1.5 m, the bottom slope should not exceed 1:12 horizontal. Where the water depth exceeds 1.5 m, the bottom slope should not exceed 1:3 horizontal.

Deck areas should accommodate chaise lounges and upright chairs and tables, the number of which should be based on facility size, market demand and the food and beverage concept. Provide a minimum 2 m deck around the perimeter of the pool with a total area of 0.25 m^2 per square meter of water surface. Safety is of prime importance in the design of swimming pools and the surrounding areas. Pools and pool decks should be designed in accordance with the US

National Spa and Pool Institute (NSPI) guidelines or equivalent standard, and all local codes and regulations.

Pool structure should be designed to withstand all anticipated loading for both pool empty and pool full conditions. A hydrostatic relief valve and/or a suitable underdrain system should be provided for in-ground pools. The designing architect or engineer should be responsible for ensuring the stability of the pool design for both full and empty conditions.

Pool decks should drain away from the pool with a minimum slope of 1%. Continuous deck drains should be provided.

6.1.2 Materials, Construction and Finishes

Swimming pools should be constructed of materials which are inert, stable, non-toxic, watertight and enduring. The building housing an indoor pool should withstand a warm, humid and aggressive internal environment and external temperatures expected in the summer season.

- Pool bottom and sides must be white or a light color, with a smooth and easily cleanable surface. The finish surface of the bottom in shallow areas 1.5 m or less in depth shall be slip- resistant (minimum coefficient of friction of 0.7 – 0.84 per ASTM C 1028.0R DIN 51097).
- Permanent depth markers are required at all pools indicating water depth in feet and centimeters, at the shallow end, all points of slope change, the deep end and at every 30 cm of depth increment. Letters should be minimum 10 cm high in a contrasting color to the coping and/or tile. Incorporate "No Diving" markers adjacent to each depth marking. Markings should be located such that they can be read from the water side and from the surrounding pool deck.
- The pool curb and scum gutter should be pre-cast masonry with a permanent white integral color and with a non-slip finish. Pool deck surface should have materials/pattern/ colors that contributes to the location of the facility site. Suggested materials are stamped concrete, natural stone, or a topical coating that simulates natural stone.
- Acoustical control should be provided for indoor pools. It is essential that advice is sought from a specialist acoustic engineer. Surface material and furnishings used for acoustical control should be cleanable and constructed of non-absorbent, water-resistant material. The design shall provide a minimum S.T.C. rating of 51 between indoor swimming pools and adjoining spaces.
- The overall building construction should provide control of the reverberation time (RT) to between 1.5 and 2.0 seconds at 500 Hz. Noise from building services (HVAC and electrical) and external break-in noise should be limited to NR 40.

Terraces and walks around the outdoor pool should be paved with an approved non-slip material, with a minimum coefficient of friction of 0.44 per ASTM C 1028.0R DIN 5109, and designed to reduce surface temperature of the paving in direct sunlight.

(a) **Indoor Pool Roof Enclosure and Ceiling:**

The roof construction / ceiling finish should be designed to following requirements:

- Provide high levels of insulation and to prevent condensation
- Be resistant to the pool environment
- Provide sound absorption
- Provide a good surface reflector for the spread of light.

(b) **Indoor Pool Doors:**

Internal door construction should be solid-core, rot- protected, water and boil proof grade doors with plastic laminate facing, lipped on all sides, suitable for a swimming pool environment. Door frames should be heavily protected, particularly the end grain near to or in contact with the floor.

Hinges and screw fixings should be powder-coated corrosion resistant stainless steel.

(c) **Pool Copings:**

Pool copings should be solid precast concrete copings with a minimum coefficient of friction of 0.44 per ASTM C 1028.0R DIN 5109, when wet.

(d) **Ladders:**

- Recessed steps, ladders, or stairs should be provided at the shallow end. Ladders or recessed steps should be provided at the deep end. If the pool is over 9 m wide, such steps, ladders, or stairs should be installed on each side.
- Pool ladders should be corrosion-resistant, and should be equipped with slip-resistant treads.
- All ladders should be so designed as to provide a handhold.
- There should be a clearance of not more than 15 cm nor less than 7.5 cm between any ladder and pool wall.
- Treads should be no more than 30 cm apart.
- Recessed steps should be readily cleanable, slip-resistant, and should be arranged to drain into the pool.
- Recessed steps should have a minimum tread of 13 cm and a minimum width of 36 cm.
- Steps should be no more than 30 cm apart.
- Where recessed steps or ladders are provided, there should be a handrail at the top of each side thereof, extending over the coping or edge of the deck.
- Where stairs are provided, they should be located diagonally in a corner of the pool or be recessed.
- Handrails should be provided at stairs such that all stair areas are within reach of a handrail. Stairs should have slip- resistant finish, a minimum tread of 30 cm, and a maximum rise of 25 cm.

(e) **Lighting:**

- Provide decorative and general lighting for the entire outdoor pool area to enhance safety and security issues.
- Artificial lighting should be provided at all indoor swimming pools which do not have adequate natural lighting, so that all portions of the pool, including the bottom, may be readily seen without glare. Lighting should be installed to provide uniform distribution of illumination.
- Overhead illumination on the water surface should be a minimum of 320 lux when underwater lighting as specified below is provided. Without

underwater lighting, a minimum illumination of 540 lux on the water surface should be provided.

- When underwater lighting is provided, at least 60 lamp lumens per 930 cm² of pool surface for outdoor swimming pools and 100 lamp lumens per 930 cm² of pool surface for indoor swimming pools should be provided.
- The design team shall verify required light levels with local ordinances.

(f) **Power Requirements:**

All electrical outlets in the immediate vicinity of the swimming pool (within 12 m of the swimming pool edge) should be protected by residual current devices and/or ground fault interrupters.

Miscellaneous:

(g) **Safety Equipment:**

Provide pool safety equipment and signage as required by local codes and regulations. Safety equipment should be posted for emergency use only.

(h) **Life Guard Stations:**

A lifeguard chair should be provided for each 186 m² of water surface area.

(i) **HVAC:**

- Indoor pools should be provided with dehumidification control to prevent condensation always on walls and glazing.
- Indoor pool room ventilation should prevent direct drafts on swimmers and should minimize condensation damage.
- See Engineering Design Criteria

(j) **Pool Drains:**

Swimming pools should be equipped with two main drains with anti-vortex covers.

6.1.3 Pool Snack Bar

A pool snack bar should provide food and beverage service. The bar counters should be designed to allow a proportion of guests to sit or stand at the snack bar, or pick up orders themselves. A covered area for table service, protected from the sun, wind and rain should be provided.

Design.

The Pool Snack Bar should be designed in close coordination with the interior designer. Due to their exposure to the elements, great care should be exercised in the design, detailing and selection of materials, to minimize maintenance problems.

Materials, Construction and Finishes

(a) **Finishes.**

Snack Bar tops and Snack Bar counter tops should be of solid mahogany, teak or other hard wood.

(b) **Floor:**

Floors of service areas should be quarry tile, sloped to drains, with removable hardwood duck boards in service work areas.

(c) **Doors and Hardware**

- Service entrance door to bar service area should be hollow metal with an approved exterior paint finish, lockset keyed under a master key and grandmaster key.
- Door should be weather-stripped, with a weather-strip threshold.
- Door should not exceed 925mm in width.
- Doors between snack bar and service work areas and the adjacent enclosed service space should be solid core doors in finish coordinated with the interior designer, with a vision panel, push-pull and automatic closer.

(d) **Electrical:**

- Provide 13A 220V convenience outlets for cleaning, point of sale terminals and kitchen printers.
- Provide work lights under snack bars and over kitchen equipment of snack bar, designed to avoid disturbance of any special decorative night lighting effects included in the overall design of the Pool Area.
- Provide power connections to refrigerators, bottle coolers, and other snack bar and kitchen equipment.

6.1.4 Pool Locker Rooms

Pool locker room facility should be provided for use by all registered fitness center members.

Planning

- The functions of control and towel issue should be combined, and so located that one person can supervise both the entrance to the pool area and the doors to the pool lockers and toilets. In addition to the lockers, this room should contain several booths for changing.
- Pool area visitors should be able to use the toilet facilities without leaving the area, and without passing through the locker and dressing rooms.

Materials, Construction and Finishes

(a) **Toilets and Showers**

Basic requirements should follow those for public washrooms, Section 6.7.2, except as follows:

- Floor of women's make-up room should be vinyl or ceramic tile in lieu of carpet, and walls glazed tile instead of wall- covering.
- The toilet partition stiles and dividing screens should be suspended hollow metal with a baked-on enamel finish.
- Separate shower rooms should be provided, with shower compartments. Each shower compartment should consist of a shower

section and a dressing section, each 90-cm square, in the clear, constructed of masonry, with a pre-cast terrazzo or ceramic tile-finished receptor, and glazed tile walls, with plastic laminate covered door to the compartment, and shower curtain between sections. Room outside compartments should be finished as specified for public washrooms.

- ***Locker Room***: Same finishes as washrooms.

(b) **Lighting and Power**

Type: LED Luminaires and/or energy saving long life lamps.

Level: Provide general lighting at a level of 216 lux with 400 lux at dressing area. Provide 13A 220V convenience outlets similar to Public Washrooms, as described in section 6.7.2.

6.1.5 Special Requirements

- Provide steel lockers 38cm x 53cm in color selected by interior designer.
- Dressing cubicles should be 1.00 m square in the clear, with plastic laminate covered partitions and doors, chrome hardware, including two double robe hooks, and an upholstered seat.
- Provide mirror with plastic laminate dressing shelf at entrance to dressing room.

6.2.0 SPA

6.2.1 General

The Spa should provide a sequence of bathing, combining heat treatments, such as hot, dry and warm wet areas, within attractively finished rooms with specialist baths and relaxation area. Treatments should be interspersed with invigorating cooling showers and relaxation that will allow bathers to revel in a sense of wellbeing.

6.2.2 Multi – Purpose Therapy Rooms

The emphasis on the treatments should be relaxation, stress management, anti-ageing and pampering, featuring a wider choice of treatments than traditionally found in salons. The therapy area should be composed of treatment rooms and relaxation rooms. The therapies carried out shall encompass a wide spectrum of beauty therapy treatments.

The therapy rooms should be designed to provide a soothing and pampering environment. Relaxing and tranquil colors, finishes, lighting and furniture are to be selected to provide a soothing and pampering environment. The therapy rooms should be large enough to accommodate rooms in various configurations of wet and dry treatments.

The designer should create moods to compliment the treatments by the use of background music, lighting scene settings, and aroma from essences. Through the application of relevant material, the designer should create a peaceful environment, free from noise from adjacent rooms, pumps, ambient noise, passing trollies etc.

6.2.3 Relaxation Lounge

The relaxation room should provide an extension to the treatment experience, by encouraging the visitors to relax between therapy treatments. The area should be configured to provide seating for 20 persons in differing size groupings. Lounge style chairs (Chaise Lounge) in differing configuration should be provided.

Provision should be made to serve light Spa beverages such as fruit and vegetable infused waters.

Locker Rooms: Locker rooms common to both the Fitness Centre and Spa should be a feature. Comfortable, spacious and fully serviced.

- The locker rooms should be divided into a wet and dry area. It is essential to have a clear delineation between the wet and dry area. The walk-through shower area should lead into the wet area consisting of sauna/steam room.
- The locker room should have a minimum of four changing cubicles with bench seating, narrow shelf with mirror above, and door. A vanity counter with mirrors and appropriate number of seated hair drier stations should be provided.

  o Minimum Requirements:

- The minimum ceiling height should be 245 cm (305 cm preferred).
- All public areas of the fitness center and Spa (doors, showers, Washrooms, lockers, telephone, water fountains, etc.) should be accessible as required by ADA, DDA regulations, and local codes and ordinances.
- Provide acoustical separation between different functional areas.
- The designer should ensure that exterior/interior doors are wide enough for accessibility and moving large equipment.

6.2.4 Materials, Construction and Finishes

(a) **Floors:**

- ***Locker room, office, Treatment Rooms***: Composite products such as Recycled rubber floor tiles complying with ASTM Sustainability Assessment.
- ***Washrooms/Showers and Steam Room***: 10 cm - 30 cm ceramic tile with minimum coefficient of friction of 0.64 per ASTM C 1028.0R DIN 51097.
- ***Sauna***: Sealed concrete to receive removable cedar duck board.
- ***Storage***: Vinyl tile.
- ***Gymnasium***: Antimicrobial carpet (should pass the American Association of Textile Colorists and Chemists test (AATCC 100 - 2012) for antimicrobial effectiveness) at exercise equipment. Provide cushioned flooring at free weights area.
- Timber floor (sports specification) in aerobics studios. The wood species selected shall be tough, resilient and long- lasting (ex. maple, pecan, red oak). The wood flooring shall be installed over suspended systems, which absorb stress while remaining strong.

(b) **Base:**

Provide 10 cm cove vinyl or rubber base. Install wood or rubber base below full height mirrors

(c) **Wall**

Locker Room, Office, Treatment Rooms: 24 oz. fabric backed vinyl wallcovering, "Plexture" coating finish or similar.

Washrooms/Showers and Steam Room: 10 cm ceramic tile.

Sauna: Cedar as provided by the manufacturer; provide glass in cedar frames at the wall to the corridor.

(d) **Storage:**

Paint on gypsum wallboard or concrete block.

(e) **Gymnasium:**

One wall should have full height 245 cm tempered or safety glass mirrors; the remaining walls should have "Plexture" coating finish or similar.

Perimeter walls adjoining any public areas (except swimming pools, locker rooms, etc.) should have a minimum S.T.C. rating of 51.

(f) **Ceiling:**

Locker room, office, massage, storage, gymnasium: Mold and Moisture Resistant Gypsum Board

Toilet/shower rooms: Painted suspended Mold and Moisture Resistant Gypsum Board.

Steam room: 10 cm ceramic tile, sloped as required to prevent dripping of condensation.

Sauna: Cedar, as provided by the manufacturer.

Doors:

Locker room, treatment room, storage: Solid core wood door with stained wood grain finish.

Gymnasium: Acoustic aluminium or wood frame door with glass.

Steam Room: Aluminium frame door with glass.

Sauna: Wood frame door with glass as provided by the manufacturer.

(g) **Lighting**

Type:

Locker room, gymnasium, storage, office, and Treatment Rooms:

LED Luminaires and/or energy saving long life lamp down lighters to cover whole room evenly.

Toilet Room: See Public Washrooms.

Shower Rooms: Recessed LED luminaire or energy saving long life downlights. Provide moisture resistant fixtures at wet areas.

Sauna: As provided by the manufacturer.

Steam Room: Energy saving long life lighting rated for use in steam room applications.

Level: See Engineering Design Criteria.

6.3.0 Outdoor Tennis Courts

6.3.1 Toilets and Lockers

The recreation facilities should be so planned that a single installation of washrooms and lockers will serve both the tennis courts and the outdoor swimming pool.

6.3.2 Materials, Construction and Finishes

- Playing surface should be cushion type composition material such as Plexipave/ Plexicushion or similar.
- All line markings should be acrylic water base paint
- Fence enclosure should be of the vinyl-clad galvanized wire fencing. Where necessary, removable or permanent wind screening arrangements should be provided.
- Lighting and Power
    - Type: 1000-watt, Metal Halide, Multi-Tap tennis court lighting fixtures.
    - Lights should be mounted 6.5 m above the court and tilted at a 45° angle.
    - Each light post should be constructed of aluminium, light in weight, capable of withstanding 100 miles per hour wind.
    - Level: Uniform 200 lux lighting level throughout the court.
    - Provide weatherproof convenience outlets as required

6.3.4 Public Area Interior Signage

The following facilities shall be Included:

- Public Area Directional Signage.
- Lift Graphics.
- Food facilities Identification.
- Exhibition and Conference facilities.
- Stairwell Signage

The specialist Interior Designer shall provide a cohesive signage package coordinated with the interior design to provide different types of signs that convey messages, directions, advertising, calls to action, or simply basic information as to the status of something.

Dynamic Digital Signage (DDS) shall be provided for:

- Advertising
- Banks
- Exhibits & Conventions
- Fitness and Entertainment
- Retail
- Food Facilities
- Transportation
- Wayfinding

The Owner's IT Department shall work with a knowledgeable digital signage system integrator to head off IT-centric issues. The location of flat panel displays shall be carefully selected to allow locating digital signage content players where they're needed, and permit integrators to plan cable runs.

- Displays shall be designed for service and maximum up time, capable of running 24/7.
- Displays shall have embedded Software inside a display or Bundled software from the display manufacturer
- Hardware shall consist of Built in processors, Video walls, High Brightness LCD etc.
- Connectivity for DDS shall be provided through wired, wireless or cellular facilities.
- The designer shall provide content that engages an audience and delivers fresh, pertinent information in real time.

SECTION 7.0

Commercial Offices

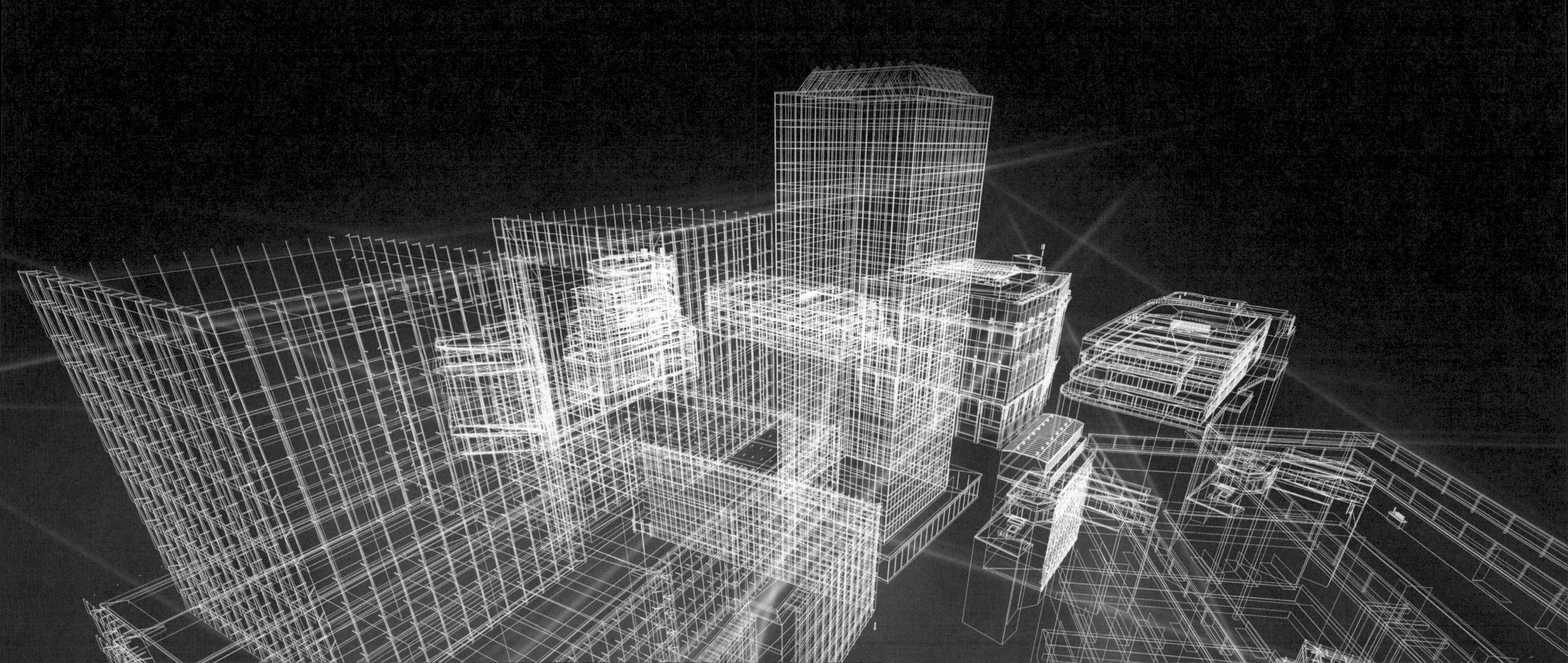

7.1.1 General

The office building must have flexible and technologically- advanced working environments that are safe, healthy, comfortable, durable, aesthetically-pleasing, and accessible. It must be able to accommodate the specific space and equipment needs of the tenant. Special attention should be made to the selection of interior finishes and art installations, particularly in entry spaces, conference rooms and other areas with public access.

Compared with major capital assets such as buildings, the fit-out of the office workplace has a shorter life cycle. Whereas major assets have life spans of 30 or more years before refurbishment or adaptation, office fit-outs have a shorter physical life of up to 15 years, but are likely to be adapted and reconfigured several times during that period. It is therefore important to ensure that, not only is the initial fit-out investment appropriate functionally and financially, but also that the fit-out is designed and constructed for adaptability and functional change in the most cost-effective way.

The proportion of the available area that is to be allocated to support spaces shall be considered to achieve an appropriate workplace density target. In some cases, a trade off might be necessary between personal spaces and support spaces to achieve the required functionality and still meet appropriate workplace density targets (12 m^2 per person).

The office building incorporates several space types to meet the needs of staff and visitors. These will include:

- Offices: Offices may be private or semi-private acoustically and/or visually.
- Conference Rooms
- Employee/Visitor Support Spaces
- Convenience Store, Kiosk, or Vending Machines
- Lobby: Central location for building directory, schedules, and general information
- Atria or Common Space: Informal, multi-purpose recreation and social gathering space
- Cafeteria or Dining Hall
- Private Washrooms
- Interior or Surface Parking Areas
- Administrative Support Spaces
- Administrative Offices: May be private or semi-private acoustically and/ or visually.
- Operation and Maintenance Spaces
- General Storage: For items, such as stationery, equipment, and instructional materials.
- Food Preparation Area or Kitchen
- Information Technology (IT) Closets.
- Maintenance Closets

7.1.2 Interior Architecture

Appropriate design principles should be applied to the design of office fit-out and several design strategies that support these principles. The design principles and supporting strategies include:

I. ***Design principle (a):***

- Design for standardization not customization
- Supporting strategies: Incorporate generic planning, modular space standards, generic workstation footprints and generic furniture profiles, finishes and characteristics.
- Plan layouts for consistency with a building's structural grid and the modular dimensions of ceilings and facades.
- Move people not fit-out to address change.
- Move electronically, not physically when feasible.

II. ***Design principle (b):***

- Design for connectivity not integration
- Supporting strategies: Design fit-out elements as separate layers that interconnect and can be disconnected and replaced/upgraded. For example, technology and communications systems should be separable from furniture systems, and visual and acoustic screening should be separable from furniture and technology systems.
- Standardize workstations and room sizes, so that work spaces and meeting spaces have the same modular sizes. This would enable changes, if required, to be economical and less disruptive.

III. ***Design principle (c):***

- Optimize hubs, nodes and zones
- Supporting strategies: Plan support functions (such as storage, meeting spaces, etc.) as hubs or nodes to increase planning efficiency and encourage social interaction.
- Introduce 'soft facilities' (such as informal seating) in the form of nodes to encourage informal interaction, networking, sharing and earning.
- Plan layouts in functional zones.
- Restrict built fit-out (partitioning extending to the ceiling) to the zone adjacent to the building's corc.
- Use the building's perimeter zone for open plan areas to maximize daylight and outlook.
- Provide alternative work spaces for staff to accommodate different types of working and varying work styles. For example, provision of space to better support collaboration such as breakout space and team tables and space for concentrated work such as quite zones located away from the communal areas.
- The introduction of a variety of alternate work settings in the workplace will better support the diversity of modern work practices and creates for a more flexible workplace.

IV. ***Design principle (d):***

- Optimize multipurpose space usage
- Supporting strategies: Design spaces and rooms to support and/or adapt to multiple uses.
- Incorporate mobile furniture and equipment.

V. ***Design principle (e):***

- Design for minimized impact on a building's structure, finishes and services.
- Supporting strategies: Avoid facilities and functions that are inappropriate for office buildings such as printing shops, photographic darkrooms and archival storage.
- Design fit-outs which are within a building's design floor loading, electricity capacity, cooling capacity and cabling capacity.
- Avoid functions or processes that affect a building's classification or compromise safety systems.
- Minimize built-in furniture and equipment that is fixed to floors, ceilings, core walls and external walls.

VI. ***Design principle (f):***

- Supporting strategies: Design to meet legislative obligations such as workplace health and safety and accessibility.
- Design for consistency with guidelines and benchmarks.
- Design for best practice.

VII. ***Design principle (g):***

- Supporting strategies: Design for ecological sustainability (community, energy, material, water).
- Design for organizational sustainability (cost effectiveness, culture and values).
- The flexible office space shall result in less refurbishment work when changes occur. The office building may be designed to be so flexible that they can also accommodate completely different uses.
- Incorporate and formalize ecologically sustainable practices for fit-out maintenance and fit-out in use (e.g. energy management, waste management, sustainable cleaning products and procedures and sustainable maintenance practices).

Enclosed offices:

- The number of enclosed offices should be minimized to maximize the options for adapting the office layout to new ways of working in future.
- Enclosed office sizes shall be based on a module of 1200 mm x 1200 mm which is the standard ceiling grid module in most office buildings. In the case of non-standard ceiling grids, the nearest modular area to those above should be used.
- Individual offices should be in the built zone adjacent to the building core to preserve the outlook and natural light for other employees. It may be appropriate to use glazed partitioning to maximize daylight, and outlook for individual offices and to facilitate effective staff supervision.
- Furniture in enclosed offices should be consistent with the modular design of workstations. Custom-designed and built- in furniture should be avoided in individual offices.

Workstations:

- Maximum flexibility shall be achieved by using generic workstation sizes configured in varying open-plan group layouts to suit teams and functions. Separating groups of workstations with enclosed offices generally should be avoided because it can constrain future changes to the sizes of team groupings and create physical barriers to effective communication between and within groups.
- Workstation systems should consist of separable components that can be reconfigured and reused without requiring multiple trades to disconnect and reconnect services.
- Soft-wired workstations made up of separate free-standing components are preferable to integrated, panel-based systems furniture that involves significant disruption whenever workstations need to be rearranged.
- Workstation footprints should be modular, with the least number of different variations possible, to allow maximum reuse of components and maximum planning efficiency.
- Sizing workstations according to function rather than classification is appropriate.
- Workstations should be based on standard ⅓L⅓ or ⅓U⅓ shaped footprints, configured in efficient clusters.
- Workstation modular footprint dimensions should be limited to: 1800 mm x 1800 mm, 1800 mm x 2100 mm, 2100 mm x 2100 mm, and 2100 mm x 2700 mm.
- Enclosure of workstations shall be provided using modular, free-standing screens individually or by group. As screen height is increased, privacy increases but communication and outlook is constrained. Screens should generally be as low as is practical, but of sufficient height to accommodate screen-based storage if needed in specific cases.

Support spaces

Sufficient support spaces should be provided to meet operational requirements, but also need to be optimized in number to avoid underutilization of space.

Multi-purpose support spaces should be used to avoid duplication and/or infrequent use.

Support spaces include:

- Meeting, interview, conference, consultation and training rooms
- Reception and waiting areas and display areas
- Registries and customer service areas
- Operational (non-archival) reference libraries
- Storage, filing and mail processing areas
- Special purpose areas needed specifically for non-standard functions
- Innovative areas for social interaction and information sharing.
- Conference and meeting rooms - These spaces should be designed as multiples of a standard 'base' module to enhance flexibility. Large conference rooms should be able to be subdivided into smaller spaces for alternative uses. Conference facilities should be shared within agencies and if possible, among agencies.
- Small Meeting Room: An enclosed meeting space for two to four persons equipped with technology enabling conference calls and/or video conferencing.

- Large Meeting Room: An enclosed meeting space for five to twelve people equipped with technology enabling presentations, conference calls and/or video conferencing.
- Unless staff training is undertaken on a day-to-day basis, consideration should be given to the use of external training facilities as needed. When training facilities are provided as part of office accommodation, they shall be shared, and should also be multi-purpose to maximize utilization.
- Print and Copy Area - Provide an open or enclosed support space with facilities for printing, scanning and copying. The minimum print and copy areas shall be one per floor or one per 50 workstations, depending on the speed and capacity of the copier.
- Reception and waiting areas - Each reception area shall be distinctive, and reflect the needs and personality of the company. These areas should be compact, functional and shared whenever possible. Because there are no typical reception layouts, room sizes, or shapes the interior designer shall layout each reception area to fit the needs and desires of the client, space plan, design and the building within which the room is to be located. As a point of entry, the reception room shall be conveniently located for direct access by both visitors and staff.
- Registries and customer service areas - These areas need to be designed to incorporate and/or adapt to new ways of service delivery and new technology.
- Storage - Storage can be classified into active, intermediate and archival types. Active (or operational) storage is associated with workstation activities and needs to be readily accessible to the user. In this case, access is frequent and forms part of the workflow. Intermediate storage refers to material that needs to be generally available but is not necessarily part of the current work process. Intermediate storage can be more centralized. Archival storage refers to high-density storage that is needed infrequently. Intermediate storage materials progressively become archival. Based on the culture of the organization, use of powerful and user-friendly tools for scanning, storing and processing data, digitization of data will eliminate the need for storing physical files and documents.
- Innovative areas - These areas are evolving as part of new ways of working. They are sometimes designed to channel people to meeting points where ideas can be exchanged in passing. Professional and expert advice is required in providing and designing these spaces to ensure that their cost is justifiable in terms of organizational value.
- Technology Room - The specialist IT Consultant shall plan this room for size, final layout, and power requirement. For initial planning purposes, the architect may take an average room size of 305 cm x 460 cm regardless of the size of company, unless the company is a trading or technology firm or has some other extensive computer usage. This room is generally centrally located within the space. When there is more than one floor occupied by the same tenant, a smaller room or large closet shall be stacked directly above or below the main technology room to allow for vertical distribution of the cables.
- Circulation space - Circulation space shall consist of primary and secondary circulation areas. Circulation space must be consistent with fire safety legislation and anti-discrimination legislation. IBC

compliance requires that the designed paths of travel for fire safety must be maintained in the approved condition and configuration. Any change to the office layout that affects the designated fire safety circulation must be resubmitted for local Fire Department approval. Expert advice must be obtained before any changes to fire safety paths of travel are made. It is not possible to identify circulation space as a definitive percentage of the total office area because of variations in building floor plates and restrictions such as structural columns and walls. A generally accepted percentage for total circulation space is approximately 30% to 35% of the net lettable area (NLA).

- Employee Facilities: Pantry Area - Specialty areas that may require a Pantry include the reception and the main conference areas. Pantries involving wet points (water supply and drainage) should be restricted to servicing board/ conference rooms directly associated with senior executive areas or to servicing lunchrooms.
- Lunchrooms - Dedicated lunchrooms equipped for sitting, relaxing, and eating on the spot shall be provided with: A sink with draining board and reticulated hot and cold water, cupboards for storage of cutlery/ crockery, chairs with back support and tables, an upright refrigerator to suit the capacity of staff in the tenancy, microwave
- Employee Lounge - There shall be two separate rooms (male and female), included with soft seating, coffee and end tables. A minimum of one lounge per 100 workstations shall be required.

7.1.3 Materials, Construction and Finishes

(a) **Floors:**

- When choosing, floor covering, the architect shall consider the environmental aspects of both the components of the product and how it is laid and maintained, and of recycling.
- All concrete floors shall be tested for moisture content (ASTM F 2170), prior to laying moisture sensitive floor covering such as resilient floor tiles.
- As required by the IBC, structure-borne noise such as footfall and objects dropped on the floor, shall be controlled to have an Impact Insulation Class (IIC) rating of 60 or higher, by applying sound deadening material to decouple the finished floor from the concrete subfloor.
- Where stone flooring is required, provide marble, granite, natural stone, or in-situ terrazzo. Polished stone or other polished materials shall have a coefficient of friction of 0.64 or greater per ASTM C 1028.0R DIN 51097.
- Resilient flooring such as linoleum, vinyl sheet, luxury solid vinyl flooring, vinyl composition tile could be successfully used in commercial office spaces.
- Carpets used in private office spaces and meeting rooms should be contract grade Nylon 6.6 or Nylon 6, with an Average Pile Yarn Density (APYD) between 4200 and 8000, and an attached cushioned polyurethane backing. A polypropylene backing should be required in high humidity areas. Preference shall be given to carpets which can be recycled, or which contain recycled fibers. All carpeting, adhesives and

cushions should be compliant with the Carpet and Rug Institute (CRI) Indoor Air Quality Testing Green Label Plus Program.

- Floors at wet locations should be resilient flooring such as vinyl tiles or linoleum tile/sheeting with a coefficient of friction of 0.64 or greater per ASTM C 1028.0R DIN 51097.

(b) **Base:**

- A wood base is required at all carpeted areas. A 10-cm wood base is preferred.
- Carpet base should be a solid color to match the predominant background color of the floor carpet.
- A wood base is required at all wood floors.
- At ceramic or stone tile flooring, a base of a matching material or wood is required.
- A vinyl base is required where resilient flooring has been applied.
- At ceramic or stone tile flooring, a base of a matching material or wood is required.

(c) **Walls:**

- Finishes should be based on Interior Designer's selection of wall coverings, cornices, and other applied finishes.
- Materials recommended for "construction" of walls or as dividers, are glass walls, framed or frameless, translucent, clear or laminated, or acrylic panels can divide spaces for privacy, yet allow light to transmit to the next space.
- Pre-fabricated full-height panels that can be installed to the ceiling and appear as a permanent wall, yet can be demounted and reinstalled as the client's needs grow and change.
- All walls receiving vinyl wallcovering should be "sized" to allow future removal of wallcovering without damaging the drywall. Wallcovering should be applied with a water based adhesive which includes a mildewcide or mildew inhibitor. All inside and outside corners should be wrapped, not cut.
- Light textured "knock-down" gypsum, or Plexture finishes are acceptable, particularly in areas of high humidity.
- Demising walls at private offices, meeting rooms, and other public areas, shall have a minimum STC 54 rating and shall extend to the underside of the structure above, complete with continuous acoustic seal.
- Any penetrations through walls with rating of STC 54, shall be sealed and caulked to maintain acoustic rating, whether the wall occurs above or below the ceiling.
- Exposed joints between door frames and walls in back of house areas shall be sealed with continuous caulking at concrete and masonry walls.
- Fire retardant wood blocking should be provided for anchoring all wall hung items such as cupboards.

(d) **Ceilings:**

- Suspended ceiling grids and dropped elements with lay-in acoustical tiles, Porous Metal tiles or similar material that support the integration of lights, diffusers, sprinkler heads, and other devices into the ceiling shall be selected.

- Conference and other types of meeting rooms require a high level of acoustical control. To achieve acoustic balance of a space, ceilings should have a NRC of 0.55 or higher. The ceiling material shall be able to selectively absorb sound reverberation and frequencies.

(e) **Doors and Frames:**

- Entry doors to private offices should be minimum 920 mm wide by 2135 mm tall.
- The architect shall provide a minimum 152 cm square area of clear floor space around each door.
- Where floor space is at a premium, sliding door may be used instead of hinged doors.
- Accessibility guidelines require that force for pushing or pulling open an interior swinging door (other than fire doors) shall be no more than 22 N for hinged doors, and a maximum of 22.2 N for sliding and folding door.
- Doors to meeting rooms and private offices should have a combined STC rating of 50.
- Where fire doors are required to be clear glass doors to support the building's functional and aesthetic needs, the door specifications shall clearly describe the fire rating of the entire door assembly.

(f) **Lighting and Power:**

- Daylighting shall be effectively integrated with the electric artificial lighting system.
- All open office areas shall be provided with strategically located optical sensors and automated controls that either turn off or dim artificial lighting in response to the available daylight in the space.
- Factors such as bi-level and multi-level switching or dimming capability as well as separate circuiting of luminaires in daylighted zones shall be considered. The control systems should be supplemented with manual override to accommodate individual differences.
- Occupancy recognition shall be packaged into the lighting control system to shut lights off when spaces are unoccupied.
- All switch plates, outlet cover plates (normal and low voltage) etc., should be designed and color coordinated with the Interior Designer in relation to final furniture layouts to assure convenient accessibility and discrete location.
- Emergency circuit fixtures should be located to give required level of illumination along exit access pathways leading to exits, exit stairs, aisles, corridors, ramps, and at the exit discharge pathways that lead to a public way (NFPA 101, Section 7.9.2.1, and IBC, Section 1006.4).
- Provide for duplex outlets conveniently located to provide power for lamps, office equipment, and housekeeping equipment. Outlet boxes should be off-set from those in adjacent rooms located in the same partition wall to prevent sound transfer between rooms.
- Lighting Levels: See Engineering Design Criteria.

(g) **Windows**

- The size, shape and location of windows should be determined by the architect in consultation with the specialist HVAC design consultant. Shading should be provided where necessary to protect from exposure to excessive sun or glare.
- The architect should ensure that the external window assembly will meet acoustic standards described in Appendix 1 - Acoustics of this manual.
- Windows should be tight without weep holes or gaps which will permit sound penetration.
- All windows should be double glazed.

SECTION 8.0

Maintenance and Engineering

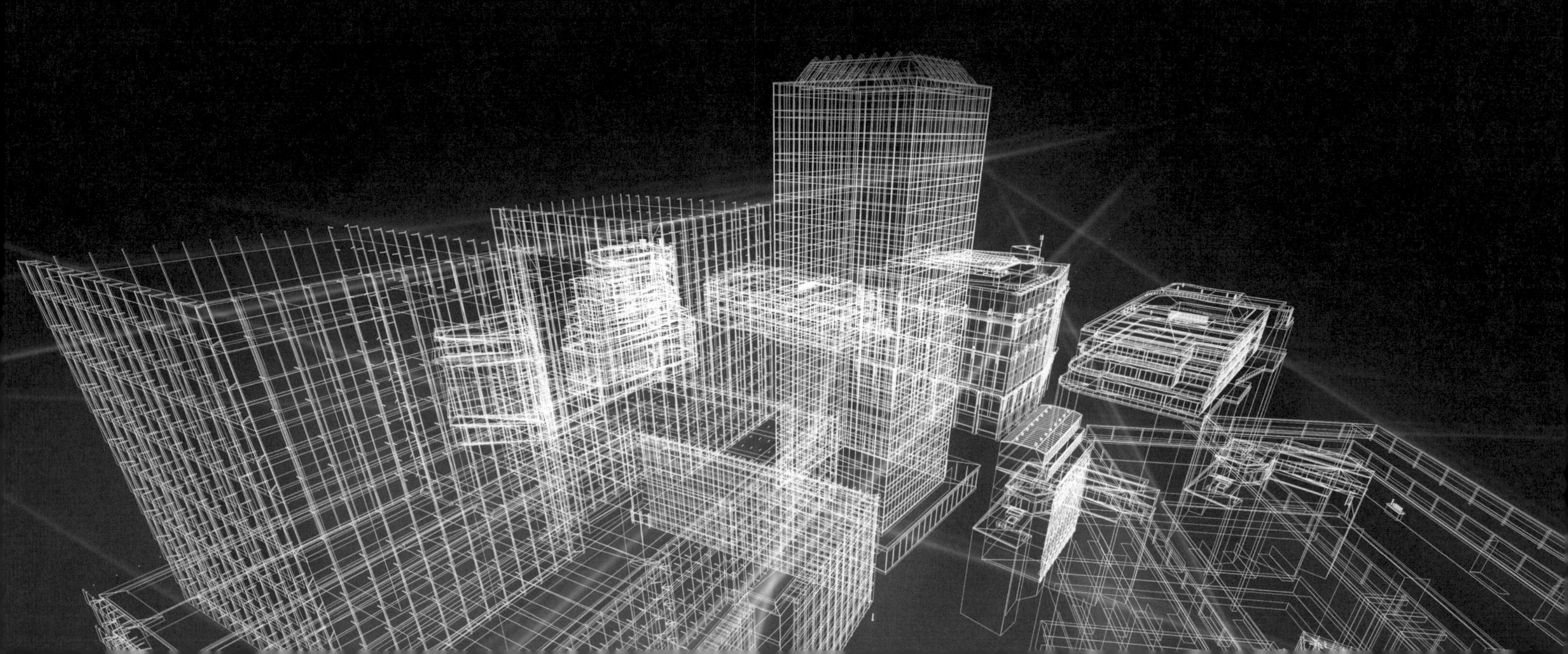

8.1.1 General

This area should include engineering administration offices, workshop facilities for repair and activities required to maintain the office block.

- ***Facilities Included***:
    - Engineering Offices
    - Repair & Maintenance Workshops
    - Engineering Store
- ***Facilities Program***:
    - Provide space for engineering services to provide proper facility maintenance and operation.

Minimum Requirements:

The Engineers Offices should be arranged so that the Engineering Manager can give as complete and continuous a visual supervision practical to the entire Engineering Department.

Facility spaces may be combined where security and safety are not critical.

8.1.2 Materials, Construction and Finishes

(a) **Floors:**

Offices: Should be vinyl tile.

All other spaces: Hardened concrete, smooth steel troweled, finished with floor enamel or a hard, dense local tile sealed concrete.

(b) **Base:**

A 10-cm cove vinyl base is required on all walls.

No base is required in rooms with concrete block walls and concrete floor.

(c) **Walls:**

Any observation windows from the Engineer's Office should be glazed with wire glass, backed with fire shutters

(d) **Ceiling:**

The plant areas shall have an unfinished concrete slab with a two-coat paint finish.

Offices should have 600 mm x 1200 mm acoustical tiles on exposed grid suspension system.

(e) **Doors:**

All doors shall be painted hollow metal or metal-clad kalamein of appropriate fire rating, and one door opening should be large enough to permit entry of all normally anticipated supplies or equipment.

Entrance doors should have half-height kick plate on bottom half of door. Upper half of door to be glass; glass panel should be as large as allowed by 2015 IBC.

f) **Lighting**

Type: T-8 Fluorescent lamps should be used with electronic ballasts. Lamp and ballast combinations should be matched.

Level: See Engineering Design Criteria.

(g) **Power Requirements**

- Duplex socket outlets should be provided at counter height and at 120 cm centers over work benches in Work Shop areas.
- Office receptacles should be mounted at 40 cm A.F.F.C.
- Provide 30A, 240/415V, 3 phase, four wire service to workshop areas.

(h) **Special Requirements:**

- The repair and maintenance workshops should have an exhaust system for airborne dust removal, and should be enclosed with two-hour fire rated partitions, or as required by latest version of IBC.
- Provide explosion proof light fixtures and a flammable liquids storage cabinet in the workshops.
- Provide a service sink in the Engineering Department.

8.1.3 Mechanical/ Electrical Room

Boilers, water chillers and fans, and all other heat, noise and vibration producers should not be located under any public space. To the extent possible they should be located on grade rather than on a structurally supported floor.

Adequate headroom shall be provided to facilitate the installation and maintenance of the large pipe, conduit and duct runs required in this and adjoining areas.

Studies should be made at the earliest practical phase of Design Development to locate main trunking and services, especially where they must pass below deep structural members.

As a minimum, two exit doorways or stairs shall be provided as egress from the Boiler and Machinery Room.

Facilities Included:

Provide a central space for HVAC, domestic water booster pumps, fire pump, hot water system and electrical systems.

Minimum Requirements:

Provide space and ceiling height as required to accommodate, maintain, service and repair specified equipment.

8.1.4 Materials, Construction and Finishes

(a) **Floors:**

Hardened concrete, smooth steel trowel finished with epoxy concrete floor paint. In a few locations, it may be economically feasible to use a hard, dense local tile which will reduce substantially future operating and maintenance costs. This possibility should always be investigated.

(b) **Base:**

Same as walls.

(c) **Walls:**

- The entire Boiler and Machinery Room should be enclosed with a 20-cm thick semi-gloss painted wall of brick or concrete masonry block to underside of structure.
- Provide a 1.80 m high wainscot of a smooth, hard surface material such as tile to facilitate cleaning by hosing down. Where tile is too expensive, provide a 1.80 m high wainscot of a hard surface enamel on the concrete and masonry walls, with white paint above. In such cases, finish of concrete and masonry work shall be specified to be smooth and straight with masonry joints struck flush and smooth.
- Provide fire rated partitions for rooms with gas fired equipment as required by the current NFPA code.
- Provide partitions with a minimum STC rating of 54 when mechanical spaces are located adjacent to public or administrative spaces.

(d) **Ceiling:**

Exposed unfinished concrete slab structure.

(e) **Doors:**

- All doors shall be hollow metal or metal-clad kalamein of appropriate fire rating, and one door opening must be large enough to permit entry of all normally anticipated supplies
- All doors should be mounted on heavy duty hinges, and in heavy bent steel plate frames, where rough usage is to be expected.
- All door openings penetrating the fire wall surrounding the Boiler and Machinery Room shall be equipped with closers and keyed under the Mechanical lock system.

(f) **Lighting:**

Type: Lighting fixtures should be industrial type T-8 fluorescent fixtures, properly located with respect to machinery and equipment layouts to provide required lighting intensity at any point in the room.

Strategically located fixtures shall be connected to an emergency lighting circuit to permit continuity of operation during a power interruption.

Level: See Engineering Design Criteria.

(g) **Power Requirements:**

Provide sufficient number of regular and heavy-duty duplex socket outlets in such locations as will facilitate the use of extension lights and electric hand tools, welding machines and other equipment for the maintenance and repair of the mechanical installation.

(h) **Special Requirements**

- Provide exterior access to this space from outside the facility, which shall be provided to permit the entry and removal of large pieces of equipment such as boilers, tanks, compressors etc.
- Provide hose bibs conveniently located for cleaning purposes, and locate floor drains at each set of equipment for emergency overflow and wash down.
- Screen view of exterior mechanical and electrical equipment from public view.
- Provide vibration isolation for all vibrating equipment.
- A small secure room or walk-in closet should be located off the general office for a key cutting machine and cabinets. This room may also contain shelving for valuable instruments.

SECTION 9.0

Engineering Design Criteria

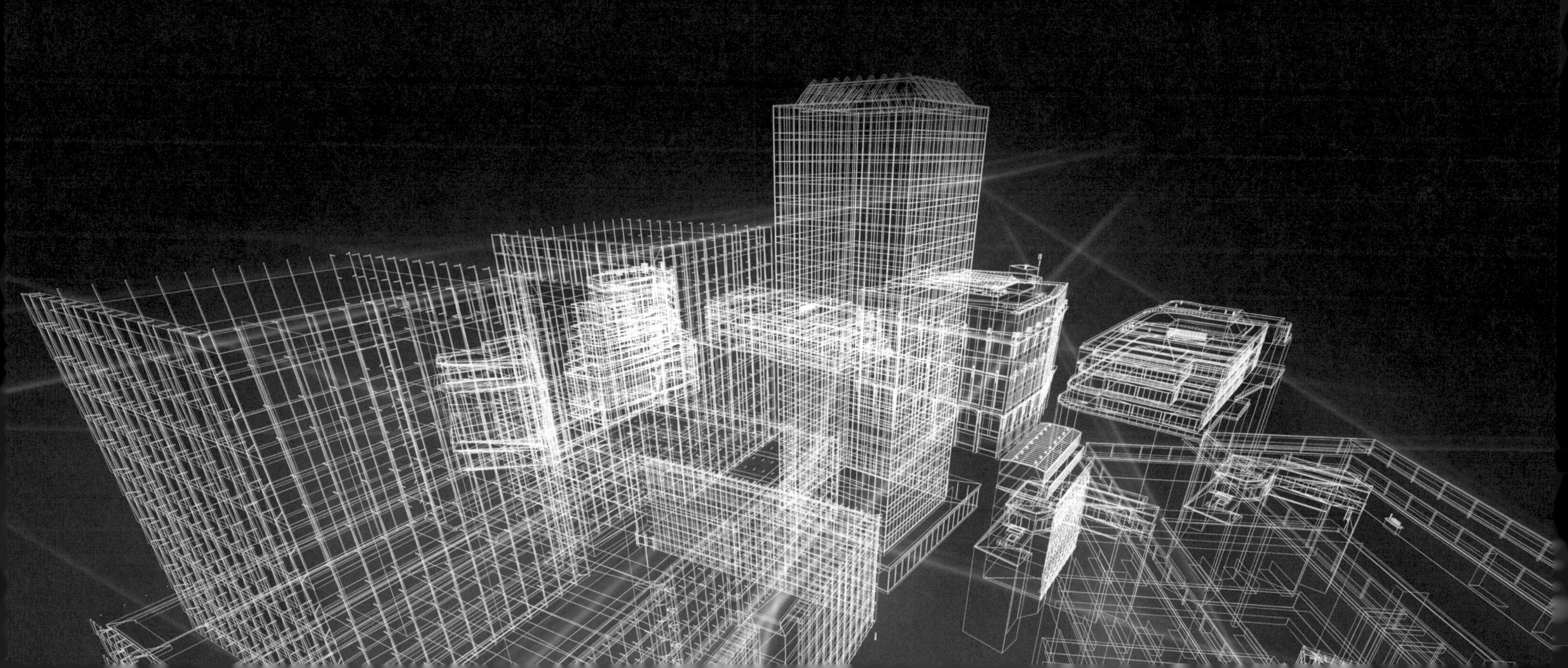

9.1.1 General

After a building's design, has been optimized in terms of envelope and daylight design, the building's performance will be determined by the efficiency the building's electrical and mechanical systems.

Several design tools and system innovations shall be applied to allow for enhanced system performance. Mechanical and electrical design engineers shall utilize computer simulation to test and evaluate projected energy use for alternative system design. Several efficiency enhancements to all components shall be considered, including premium efficiency motors, variable speed drives, direct digital control (DDC) systems, and advance control strategies. Through well integrated design, equipment may be appropriately sized [not oversized] for part load efficiency, and reduced initial cost.

The design team shall coordinate planning and design work with the architectural, structural, civil, site, and sustainable designs, as applicable.

The following section is meant as a design guide. Where the requirements of this section, and local codes conflict, the more stringent code requirement should be applied.

Special requirements appear throughout this document and are listed by facility. These requirements supplement, and in some cases, supersede, the requirements in the following section.

All equipment, components and materials should be rated and selected for the service temperatures and pressures applicable to their physical installation, including adequate pressures due to start-up, shut down, water hammer, etc.

Appropriate materials and corrosion/moisture protection should be specified for equipment being installed in corrosive atmospheres such as marine locations. Review the locations of equipment, components and materials, the type of corrosion protection, accessibility for maintenance and repair. A report on the protection to be used is required if the mixed-use building will be built within five miles of a body of salt water.

9.1.2 Design Standards

The Designer shall agree with the Owner, the standards to be adopted in the design and procurement of the design. The services should comply with recognized International codes and standards, and where there are conflicts between these standards and local standards, these are to be brought to the attention of and agreed with the Owner. The more stringent requirements should be complied with. Recognized international standards that are deemed acceptable include the latest version of:

- International Plumbing Code
- International Mechanical Code
- NFPA
- CIBSE
- ASHRAE

Where alternative standards are proposed, these should be submitted for approval, with clear indication of any deviations from any of the above standards. For consistency of design, insurance purposes, and safety / comfort of occupants, the above recognized International standards should be complied with, over and above any local standards.

The designer should be responsible for making and obtaining all local regulatory approvals. This may involve the preparation of specific documentation highlighting variances from the local standards arising from this brief.

From a design standpoint, the mechanical, electrical, plumbing (MEP), and fire protection design team shall establish the owner's intentions related to initial and potential long-term use of the building. It may be necessary to delineate the MEP infrastructure across usages to ensure that operational costs are appropriately separated.

9.1.3 Designers Responsibilities

The designers commissioned to design the mechanical and electrical systems are deemed to have understood the obligation to familiarize themselves with the relevant international standards and highlight any conflicts with local standards for early resolution with the Owner. Where any such conflicts between standards exist, these should be brought to the attention of the Owner during the design process or within sufficient time so as not affect either the design and/or the construction program. Where any doubt exists, the issues should be raised for Owner's decision or advice.

This MEP and Fire Protection services design criteria manual has a series of 'Design Stage Compliance Verification Schedules' which should be completed by the designers to verify conformance with Owner requirements. Failure to complete these schedules may result in the Owner requiring an independent audit of the design to be carried out at the designer's expense.

9.1.4 Technical Submissions

The designers are required to make technical submissions as detailed elsewhere in this manual at the various stages of the project as indicated. These submissions should be made in English language, and should be to a format to be agreed with the Owner. Where the Owner has a Technical Services Engagement for assisting the designers, these technical submissions will be reviewed and commented upon.

The process of technical submissions does not obviate the Designer's responsibility to ensure full compliance with the requirements herein. Unless variations from these requirements have been given a specific approval by the Owner, then they may not be acceptable and may require rectification at the designer's costs.

9.1.5 General Design / Installation Principles

Mixed-Use buildings are unique developments in that they could be operational for 24 hours a day, with occupants paying for comfort and life safety being an expectation. All materials and equipment used should generally have an expected life of at least 25 years, or as appropriate for a specific system, to be approved by the Owner if different from this, or in accordance with CIBSE or ASHRAE recommendations.

The focus of the engineering systems shall be towards sustainable systems that are budget-friendly with low energy consumption and short return on investment.

All services should be located in positions that are readily accessible for maintenance without causing disruption to the normal operation of the building. No materials should be used or specified which are deleterious or harmful to human health or the environment, either during their installation or during long term disposal or degradation. This includes, but is not limited to the following:

(a) **Deleterious Materials**

- Asbestos
- CFC's
- Crystalline Silica
- Formaldehyde
- Lead
- Manmade Mineral Fibres
- Polychlorinated Biphenyls (PCB's)
- Volatile Organic Compounds (VOC's)
- Wood Preservatives
- Calcium Chloride
- Calcium Silicate brickwork
- High Alumina Cement (HAC) concrete
- Wood wool slabs or wood wool cement boards

(b) **Harmful Materials**

- Cement fibre slates
- Composite panels
- Galvanized steel wall tiles
- Hollow clay pot floors
- Nickel sulphides

9.1.6 Deviations from the Owner's Standards

Where deviations from the Owners standards are agreed (in writing), the designers should maintain a schedule of the agreed variations including:

- The item reference and applicable section of the Ownber's standards
- The reason for the variation
- Who agreed the variation
- The date
- The cost/program benefit to the project

9.2.1 Utility Services

The utility services should be subject to a reliability study but general principals should be:

(a) **Electricity**

The utility company shall supply either an MV or LV electrical supply dependent on the nature of the local network availability. All incoming electrical supply transformers and switch rooms shall be located at the back of house areas of the building.

(b) **Water**

A dedicated domestic water supply shall be provided to serve all of the building 'non-fire' requirements from the Water Supply Company's mains. The Water Supply Company's mains should be fitted with a water meter at the site boundary, as well as a stop valve either side of the water meter.

A dedicated fire water supply shall be provided to serve all of the building 'fire' supplies, which may include:

- First Aid hose reels
- External fire hydrants
- Internal wet rising mains
- Sprinkler systems
- Fire water supply to sprinkler/hydrant storage tanks

The fire supply should comprise of dual incoming supply. Where the facility exists within the local water reticulation, the dual incoming supply should be derived from separate external water mains.

(c) **LP Gas**

Gas shall be from underground bulk storage vessels or a central LPG station with manifolds and distribution system. The central LPG shelter and distribution system shall fully comply with the NFPA 58 (Liquefied Petroleum Gas Code). Gas shall be provided to the site from the local service provider, and should enter the property in a position that allows unobstructed 24-hour clear access by the utility company. The gas farm assembly should be at the rear of the building, not accessible by the public unauthorized persons.

(d) **Drainage**

Drainage discharge should be to the public sewers where available. Planning conditions should be reviewed for possible requirements for surface water drainage, balancing ponds or tanks.Wherever possible, the drainage systems should be designed to operate without pumping. However, where pumping is unavoidable, the pumping stations should generally be submersible wet sumps with a minimum of 2 No. pumps and 100% redundant pumping standby facility. The pumping facility shall be fed with dual supplies from both the normal power supply and the emergency standby supply, with a local auto changeover device. Drainage should be run in direct routes, with benched manholes provided at all changes in direction.

(e) **Telephones**

Telephone and Internet access facilities shall be provided to the site from the local service provider.

9.3.1 Life Safety Systems

The objective of these systems is to protect life and property in the event of a fire or incident. This section of the manual should be read in conjunction with the Life Safety - Section 2.0 of Architectural Design Criteria.

General

The Life Safety Systems for the building include the "passive" as well as "active" design features, which should be designed in co-ordination with each other. The aim of the design of these systems is to meet or exceed the Regulatory requirements, whilst simplifying the operation and maintenance of the fire services as much as practically possible. As a minimum, NFPA standards or British Standards shall be acceptable, with the inclusion of any specific requirements noted herein.

The designer should provide a "Fire Safety Strategy" for the whole building, which should include all passive and active systems, and clearly show:

- The anticipated risk of a fire and its severity.
- The Architectural provisions for resisting the spread of fire and smoke, including provisions of occupants escape routes, identification of fire compartmentation, fire doors and shutters.
- The adequacy of services designed to detect and prevent the spread of smoke and fire – including details of the Standards/ Codes to be adopted in the design of these services.
- A Fire Detection and Alarm System Sequence Matrix – refer to Table 11.30
- The provision of facilities to assist the Fire Services in locating and fighting fires.
- Areas of refuge and Lifeboat evacuation capabilities
- Outline schematic layouts to illustrate the fire strategy.

Once this has been approved in principle by the Owner, the designer should be responsible for obtaining the approval of the Local Regulatory Authority.

9.3.2 Fire Protection Systems

(a) **Objectives**

Provide protection of life and property by providing both automated and manual facilities for preventing the spread of fire.

(b) **Design Standards**

The overall fire protection concept for the building shall be based on the following protection objectives:

- Protection of human life
- Protection of material assets
- Prevention of business interruption

Fire risks shall be defined in a multilevel fire protection concept that defines specific protection objectives. Therefore, each likely fire location shall be protected by adequate measures so that no incipient fire can grow up to a serious fire.

Which protection measures and how many of them should be implemented would require an in-depth fire risk assessment, both in evaluating the protection needs and prescribing adequate protection measures.

Also refer to section 9.1.2.

(c) **Description**

The following systems should be provided as required by the relevant codes, or to enhance the life safety provisions of the building:

- Fire Alarm and Detection
- Automatic Fire Suppression – (all areas to be covered)
- Wet Risers and Hose Reels
- Hydrants (external)
- Fire dampers between fire zones Zone
- Smoke clearance using installed air handling plant
- Automatic and manual cut off fuel/power to kitchen equipment and boilers in a fire alarm in that area.

9.3.3 Fire Alarm and Detection

(a) **Objectives**

Provide protection of life and property by the early detection automated alarm/ annunciation in the event of a fire incident to the whole building. Fire detection systems shall be required at varying levels in the nonresidential buildings, based on the building type and its contents, the occupancy, and the minimum requirements within the codes and standards.

Notification appliances are to be included in the design to ensure building occupants are notified and safely and efficiently evacuated.

(b) **Design standards**

All in accordance with section 9.1.2.

(c) **Control requirements**

An intelligent, microprocessor controlled, fully addressable, automatic fire alarm and detection system should be installed to cover the entire building. This system should have a small time delay if permitted under local regulations to enable staff to investigate the cause of an alarm. This is to reduce the level of false alarms and hence disturbance of building operation.

(d) **Description**

A software-driven fire alarm and detection system should be installed to cover the entire building. The system should comprise fire and other alarm initiating devices, general supervision detection devices, notification appliances, as well as, addressable/intelligent fire and combination alarm control systems. The system shall have the following equipment, to provide life and property protection to all areas of the property:

- Main fire alarm panel located in the Fire Command Centre with test facility.
- Repeater fire alarm and detection panel located at Security Control Room and Engineering Manager's Office.
- Addressable point smoke and heat detectors with time delay facility at the main panel. Selection of automatic detection devices shall be based on the type and size of fire to be detected and the response required.
- To detect incipient fires in high-value or critical areas, such as critical computer rooms, before flame or noticeable smoke develops, a very early warning fire detection system such as an air-sampling system, or spot-type high-sensitivity smoke detectors shall be provided. For fastest response, locate detectors or sampling points inside electrical/electronic cabinets.
- Ionization smoke detectors shall be installed to detect a flaming, well-ventilated, energetic fire, such as that involving a transformer.
- Photoelectric smoke detectors shall be installed to detect a smoldering fire that produces smoke with large, super- micron particulates, such as that involving furniture, paper products, or commodities in cardboard boxes.
- Combination ionization/photoelectric smoke detectors shall be provided to detect a slow smoldering fire, such as that in switchgear and circuit breaker rooms.
- Provide spot-type fixed-temperature heat detectors in spaces that may reach high ambient temperatures under ordinary conditions (such as boiler and electrical wiring rooms), or as required by the applicable occupancy. A heat detector with a temperature rating slightly higher than the highest expected ambient temperature, shall be provided.
- The Alarm and trouble signals from each building fire alarm control panel shall be digitally encoded by UL listed electronic devices onto a multiplexed communication system.

Initiating device circuits and signaling line circuits shall be connected to monitor any of the following:

- Manual fire alarm box
- Automatic fire detection, such as smoke, heat, flame, etc.
- Sprinkler water flow detection
- Control valves
- Pressure supervision for pressurized water storage tanks
- Water level supervision for gravity tanks, pressurized water tanks and pump suction tanks
- Automatic fire pump supervision
- Supervision of other fire suppression systems
- Interface to the Lift installation and security system
- Interface to air handling systems for engineered smoke clearance systems

Note: Any alternate positioning of either the Main or Repeater Fire Alarm panel should only be carried out with the written approval of the Owner

Detector spacing should be in accordance with the relevant statutory code or manufacturers recommendations but should be a minimum of one smoke detector per 85 m^2 or one heat detector per 50 m^2, and every 15 m along corridors (less than 2m wide). Detection is required in all storerooms over 14 m^2.

Where detectors are installed in concealed zones, risers, floor voids and ceiling voids, etc., then a remote visible indicator should be installed. These should be installed in all voids in excess of 80 mm unless there is no combustible material, or in compliance with the local regulatory code.

Where atriums are present, optical beam detectors should be provided for detection of smoke in the atrium and operation of the atrium smoke clearance system.

Smoke detectors should be provided in the supply duct for shut down of associated Air Handling Unit (AHU), and annunciation at the Fire Panel.

Manual call points should be installed as a minimum at all final exits plus emergency exits of each zone, so that the maximum travel distance should not exceed 30 m to the call point.

The fire alarm and voice communication system shall be highly flexible and scalable to increase in size as necessary, through the use of additional panels, amplifiers, and power supplies.

(e) **Emergency Communications.**

- A one-way emergency communication system for the protection of life by indicating the existence of an emergency situation and communicating information necessary to facilitate an appropriate response and action shall be installed. The design of these systems shall be based on a risk analysis that characterize the probability and potential severity of incidents associated with natural or man-made disasters or other events requiring emergency response. This process would require a threat and vulnerability survey to identify the risks.

Audio-based way finding systems shall make exit locations highly noticeable in an emergency, guiding occupants directly to a means of egress. This should be accomplished by combining audio and visual alerts.

Point voice evacuation speakers should be provided throughout the building with Xenon beacons in public corridors, plant rooms, kitchens, laundry and other areas with high ambient noise levels. Fire alarm sounders should be audible throughout the building (including residential apartments) to the required sound level mandated by the NFPA code.

It is critical that the voice notification system. Visible notification shall be provided for the hearing impaired, and Texural, graphic or video displays may be used in addition to the strobes.

The latest version of NFPA 72 chapter 24 shall be used in the design and application of in-building fire emergency voice alarm communications. Audibility shall be required in all areas of the building in accordance with Chapter 18 of the NFPA 72.

- The design shall clearly describe the management process for fire alarm notifications between different structures and occupancy uses within the same building.
- A two-way telephone communication system should be provided throughout at all entries to exit stairs and corridors for Fire Department use, from the main fire panel located in the Fire Command Center.

The communications system shall operate between the Fire Command Center and every lift car, every lift lobby,, and each floor level of exit stairs, mechanical plant rooms, lift machine rooms, and air handling (fan) rooms

(f) **Fire Alarm System Wiring**

The installation of all wiring and cable, whether actual metallic conductors or fiber optic media, should conform to applicable requirements of ANSI/NFPA 70 National Electrical Code (NEC). or applicable local codes. Enhanced fire-resistant cabling shall be installed to NFPA, British and European standards.

Article 760 of ANSI/NFPA 70 is specifically devoted to fire alarm systems, but many other code provisions directly affect the system installation, such as those for general provisions, wiring and protection, grounding, wiring methods and materials, special occupancies (such as hazardous locations), emergency systems, optical fiber cables, etc. The requirements in ANSI/NFPA 70 National Electrical Code shall be coordinated with ANSI/NFPA 72 National Fire Alarm Code.

When fire alarm signaling circuits are installed as cables (metallic or fiber optic), cabling shall be rated and listed for use in general applications, risers (vertical chases or shafts) or in plenums (air handling spaces).

Isolation modules should be provided at the entry and/ or exit from each separate area of a designed fire zone.

As minimum, isolation modules should be provided, to ensure that no more than 2000m² of covered area becomes unprotected in the event of any single cable fault and no more than 10,000m² of covered area becomes unprotected in the event of any two simultaneous cable faults. Designated loop isolators should also be provided for monitoring the zone states

Fire alarm cables shall be separated from non-fire alarm circuits or circuits of other voltages.

(g) **Storage Batteries**

Storage batteries shall be provided in the secondary supply for the fire alarm system, as an integral part of the main fire alarm panel. Proper sizing of batteries for fire alarm system is a critical factor, and the manufacturers calculation methods, tables or other techniques shall be applied to assure a minimum size battery is provided based on the system configuration, standby load and required duration, and alarm load and required duration. It is essential that the selection is made according to the manufacturer's instructions.

(h) **Fire Detection & Alarm System Sequence Matrix**

The designers are required to provide a specific "Cause and Effect" Matrix for the services incorporated into the design. The following is a "typical" example of what is required, but this should be specific for each Project.

Table 9.3 Sample Fire Detection and Alarm Cause and Effect Matrix

9.4.0 Fire Suppression

Fire suppression systems shall be designed in accordance with the International Fire Code, and the local Civil Defense specific requirements.

Complete fire suppression systems are to be provided and installed in accordance with the applicable International Building Code (IBC) and National Fire Protection Association (NFPA) standards. These shall be as follows:

- Automatic Sprinkler System
- Standpipe System
- Clean Agent Extinguishing Systems
- Cooking Equipment Suppression Systems
- Portable Fire Extinguishers

The design team shall factor-in the retail component, which generally represent the greatest fire protection water demand., beside a standpipe system. If tenant occupancy results in sales racking or storage racking within stockrooms, that are 4.5 m or higher water flow demand could be considerably higher.

9.4.1 Automatic Sprinkler System

Fire Sprinkler System design criteria shall comply with latest version of NFPA 13 – Standard for the Installation of Sprinkler Systems. Hydraulic calculations shall be used for design.

Designers should also refer to NFPA 13E, Recommended Practice for Fire Department Operations in Properties Protected by Sprinkler and Standpipe Systems.

It is essential that the sprinkler systems designers and fire alarm system designers work together in buildings of any size or complexity. The fire alarm system annunciator should indicate the location of the alarm to the local fire service. Coordination is essential to furnish the fire service with clear information on the fire or its location.

Large floor levels should be divided into zones. This should accomplish the following:

- Allows the fire alarm to indicate the fire location more specifically within a floor
- Limits the system area taken out of service for maintenance, repairs, or renovations.

Valves that control sprinkler systems or specific zones shall normally remain open, and shall be supervised electrically by the fire alarm system.

A double-interlock pre-action sprinkler system shall be provided for computer rooms, data centers etc., where water filled pipes above critical equipment is of concern. The double-inter lock system shall require both smoke/ heat detection and a loss of gas pressure in the pipe to occur prior to system activation and discharge.

9.4.2 Standpipe Systems

When required, standpipe systems must be installed in accordance with NFPA 14, Installation of Standpipe and Hose Systems. The system shall be designed by the hydraulic calculation method.

In buildings with standpipe systems, sprinkler systems are usually combined with them and fed by a single water supply. Typically, all sprinklers would be located downstream from a zone control valve that will shut off water to all sprinklers but not to fire hose connections. This will allow the hose connections to remain available for manual fire suppression during times when one or more sprinkler zones are shut off for any reason, either before or during an emergency incident.

Class I standpipe systems should be provided in exit stairways of buildings four stories or more in height.

9.4.3 Clean Agent Extinguishing Systems

Clean agent fire extinguishing systems shall be provided for risks where a dry, clean and non-toxic extinguishing agent is required that does not damage the high value content material to be protected.

- Clean agent fire extinguishing systems must conform to NFPA 2001.
- Each system shall have stand-alone (not dependent upon the building fire alarm system for operation) control panels that are listed for releasing device service, and monitored by the building fire alarm system.
- Careful consideration should be given to compartment under/over-pressurization during the discharge of total flooding clean agent systems
- A manually activated exhaust system shall be provided to facilitate the extraction of any remaining clean agent after the required hold time of the total flooding clean agent system. The exhaust system should be integrated into the HVAC system for the enclosure.

9.4.4 Cooking Equipment Suppression Systems

All commercial grease hood and ducts shall meet the requirements of NFPA 96 – Standard for Ventilation Control and Fire Protection of Commercial Cooking Operations, and NFPA 17A – Standard for Wet Chemical Extinguishing Systems (as appropriate).

Automatic wet chemical fire suppression (or water misting or other approved system) should be provided for cooking ranges in the kitchens. Activation of this system should:

- Interface with the main building fire alarm system for annunciation.
- Isolate the gas and power supplies to all kitchen equipment automatically on activation of the suppression system
- Switch off the supply and exhaust ventilation system serving the kitchen

Table 9.3 Fire Detection and Alar m Cause & Effect Matrix

		ANNUNCIATION					**SMOKE CONTROL**					LIFTS	
	System Sequence Alarm Initiation Device	Annunciation & Printing at FACP & RRP	Annunciation at FACP & RRP Graphic	Voice Evacuation	Sound Alert Tone on All Other Floors	Sound Message in emergency Floors	Start Pressurization Fans	Stop Pressurization Fans	Start Smoke Extract Fans and open dampers in Associated Areas	Stop All Supply and Return Fans Associated with the Area	Release Smoke and Fire Doors	Elevator Recall to Lobby Floor	Alternate Floor Recall (Not the Alarm Floor)
Priority 1 Alarms	Area Smoke and Heat Detector	X	X	X	X	X	X		X	X	X	X	
	Supply/Return Air Duct Detector	X	X	X	X	X	X		X	X	X	X	
	Sprinkler Flow Switch	X	X	X	X	X	X		X	X	X	X	
	Manuak Breakglass Station	X	X	X	X	X	X		X	X	X	X	
Lift Controls	Lift Lobby Smoke Detector	X	X	X	X	X	X		X	X	X	X	
	Lift Machine Room Smoke Detector	X	X	X	X	X	X		X	X	X	X	
	Lobby Floor Lift Lobby Smoke Detector	X	X	X	X	X	X		X	X	X	X	X
Secondary Systems	Kitchen Hood Suppression System	X	X										
	Special Electrical Room Protection System	X	X										
	LP Gas Leak Detection System	X	X										
	Stair Pressurization Fan Smoke Detector	X	X		X			X					
Monitoring	Sprinkler Valve Tamper Switch	X	X										
	Fire Pump Monitoring	X	X										
	Water Storage Tank Low Level	X	X										
	System/Wiring Fault	X	X										

Grease ducts shall be protected by approved products, designed with clearance reduction methods and installed as fire rated enclosures.

9.4.5 Portable Fire Extinguishers

Portable fire extinguishers are to be provided throughout the facilities based on occupancy, length of travel between extinguishers, and hazard as required per NFPA 10.

- The maximum travel distance to a fire extinguisher should not exceed 25m.
- Kitchens using deep fat fryers or other appliances utilizing combustible liquids shall have the appropriate size class K fire extinguishers located within 10 m of such appliances.

9.5.0 Emergency Lighting

9.5.1 Objectives

Emergency lighting is just one component of the means- of-egress, and part of a building's life safety system. In the event of a power outage, a backup lighting system may be the occupants' only guide to a safe exit.

9.5.2 Design Standards

NFPA 101, Section 7.9.2.1, and IBC, Section 1006.4 require that emergency lighting provide a minimum average of 1 fc for a period 1.5 hour with a minimum of 10 lux at any point, and a maximum illumination level that cannot exceed 40 times the minimum. The designer is required to provide calculations to prove these levels have been achieved.

Emergency lighting shall be connected to a secondary power source. Emergency lighting is required to be equipped with at least two sources of functional power so that in the event one source fails, it does not affect the capability of the second source. The lighting systems shall comply with the International Fire Code, NFPA 70 - National Electric Code or BS EN 60598-2-222, code of practice for emergency lighting.

The emergency lighting should have a minimum duration of 3 hours unless otherwise approved by the Owner.

Non-maintained luminaires shall be provided in areas that require emergency lighting only on mains failure. Maintained emergency lighting systems shall be provided in:

- Residential buildings
- Recreational Areas
- Non- Residential public premises such as shopping malls, retail shops, covered car parks etc.

Maintained EXIT signs shall be provided throughout the building.

9.5.3 Control Requirements

The system should be so configured to ensure that it automatically operates not only under complete failure of the main electrical system, but also under sub-main and final circuit (local) failure.

High activity areas such as gymnasiums, kitchens, etc., should be supplied with emergency lighting within half a second for 20% of the luminaries.

9.5.4 Description

The system should be a central battery inverter type system. The system should provide emergency power to the lighting installation upon power failure and support emergency exit signage, with full test facilities. Emergency lighting should provide illumination to all escape routes.

In addition, the areas identified in the schedule of emergency generator loads should be illuminated.

In addition to emergency lighting for means of escape a minimum illumination level of 10 Lux measured at floor level should also be maintained in all areas covered by security cameras.

All wiring associated with central battery systems, including the interconnection cables between the central battery, emergency lighting distribution boards and luminaires or slave conversion modules, should be carried out using fire rated cables, which should be differentiated by color from the fire alarm system cabling.

9.6.0 Smoke Control/Clearance

9.6.1 General

These systems fall into two categories: exhaust and pressurization. Each of these types of systems has its place and be a very effective means of safeguarding the building occupants from exposure to the products of combustion. The design engineer should tailor the system to the characteristics of the building and its occupants.

Buildings in excess of six stories or more than 23 m above the lowest level of fire department vehicle access should be provided with:

- Stair pressurization in fire exit staircases
- Lobby relief in public corridors (for smoke exhaust and to enable the pressurization system to operate)
- Zone smoke control to all public areas

In addition to sizing fans and associated ductwork properly for smoke-management systems, the designer shall specify appropriate temperature-rated equipment (i.e., fans, dampers, ductwork) that will be in contact with the smoke-plume/hot upper layer.

The design engineer shall apply Computational Fluid Dynamics (CFD) modeling to build a computer model of the space to simulate various fire scenarios that can be used to optimize the design. The engineer should run the model with various fire sizes and locations as well as modify the airflow rates and locations of exhaust and makeup air.

9.6.2 Smoke Control

The design of a smoke control system could be a mechanical (active) or a passive system or a combination of the two systems. The mechanical system should use either the HVAC system or dedicated smoke control fans to extract smoke from the building, or create a pressure differential to prevent the migration of smoke from the area of fire origin. The passive system will rely on the natural buoyancy of the smoke and the stack effect to allow the smoke to exit the building.

Smoke control in large open spaces, such as malls and atria should be achieved through mechanical exhaust. The design of the smoke exhaust system should be based on a thorough and rational analysis that considers the geometry of the space, the fuel load (the expected magnitude of a fire in the space), the means of introducing makeup air to replenish the volume gas being removed by the fans, and the effects of the air movement on the adjacent egress components. In addition to the objectives for an exhaust-type smoke control system to keep the smoke layer at least 1.83 m above the highest walking surface in the atrium, set by the latest version of the IBC section 909, guidance provided in NFPA 92: Standard for Smoke ControlSystems shall be applied in the design of the system. Smoke dampers used as part of the smoke control system must be ANSI/UL 555S: Standard for Smoke Dampers-listed. This standard will ensure that the dampers can withstand potentially elevated temperatures and higher pressures, and will allow minimal leakage rates through the damper.

9.6.3 Stairway Pressurization

The IBC requires all interior exit stairways serving floors more than 23 m above the lowest level of fire department vehicle access to be designed as smoke proof enclosures. The design criteria for stairway or lift hoist way pressurization systems have some of the differences between the IBC and NFPA 92. Table 1 illustrates some of the differences between the IBC and NFPA 92 that must be taken into account when designing a stair pressurization system where both documents are applicable.

9.6.4 Control systems

The designer shall apply specific criteria described in NFPA 92 and the IBC for the subsystems that control the overall smoke control system.

Control systems shall be listed in accordance with ANSI/UL 864, Standard for Control Units and Accessories for FireAlarm Systems, category UUKL.

Simplicity should be the goal of a control system design. A single system should be used to control the various smoke control functions.

Activation of the smoke control system should be via the area smoke detectors or sprinkler flow switches (where sprinkler zones match smoke zones), and should activate the zone AHU under smoke alarm to full exhaust and surroundings unit to full fresh air.

Special design consideration is required where smoke stratification can occur, such as within tall atrium spaces. Upward-facing beam-type smoke detectors or detection at multiple elevations within the space should be used.

A firefighters' smoke control station (FSCS) located adjacent to the fire alarm panel in the Fire Command Centre is required for all smoke control systems per NFPA 92 and the IBC. The FSCS shall provide manual control, status indicators, and fault conditions for the system. Manual controls should be clearly marked and should show the graphic location and function served via diagrams or notations on the FSCS. Means of verifying correct operation of components upon activation must be provided. This includes positive confirmation for the operation of fans, any fault conditions, and manual overrides. Failure to receive or maintain positive confirmation of operation must provide an off-normal indication within 200 seconds. The systems should be "hard wire" controlled direct from the fire detection and alarm system in fire rated cabling.

NFPA 92 requires that the smoke control mode must be initiated within 10 seconds after an activation command is received. For smoke containment systems, the fans must operate within 60 seconds. Completion of damper travel must occur within 75 seconds. For smoke management systems, full operational mode must be achieved before conditions exceed design smoke conditions.

9.6.5 Testing

Criteria for testing described in NFPA 92 shall be fully applied. The systems should be tested to clear "cold smoke" so that exit signs are visible within 10 minutes of activation. Smoke ventilation to basements and car parks should be provided in accordance with NFPA standards and local code requirements.

It is important to discuss the variables with the local fire department and reach a consensus early in the design process. Also, summarizing the testing procedures in the design documents for approval would avoid issues during the acceptance testing.

The designer shall provide relief dampers to ensure door- opening forces are not exceeded during testing or during a real-world scenario. Relief dampers shall be used to overcome the potential of excessive pressure build-up when a door is opened wand then closed.

9.7 LPG Storage

9.7.1 General

Mitigation measures such as space separation, and emergency isolation systems shall be provided, where accidental release of LPG will expose high-value property. LPG storage and distribution system shall be designed in compliance with NFPA 58 or a similar international standard.

9.7.2 Storage Tanks

LPG storage tanks shall be located 61 m from main buildings. If a 61m separation distance is not feasible due to site constraints, provide the distances as shown in Table9.4.

Table 9.4 Separation Distances to Main Buildings for Aboveground Stationary LPG Tanks

Tank Capacity (m^3)	Single Tank (m)	Tank Group (m)
1.9 – 7.6	7.5	Aggregate Capacity Not Exceeding 7.6 m^3
7.6 – 230	23	Aggregate Capacity Not Exceeding 680 m^3
230 – 340	46	

9.7.3 Emergency Isolation and Detection

- An accessible, manual shutoff valve shall be provided externally, where LPG piping enters a main building and at each gas-consuming unit to permit access in the event of fire at the unit.
- LP gas detection should be installed where a gas intake point is such that any escape could have serious consequences. Interface with the fire alarm & detection system and fire suppression system should be provided to shut down the gas supply on an alarm.

Aboveground Stationary LPG Tanks

SECTION 10.0

Mechanical/HVAC

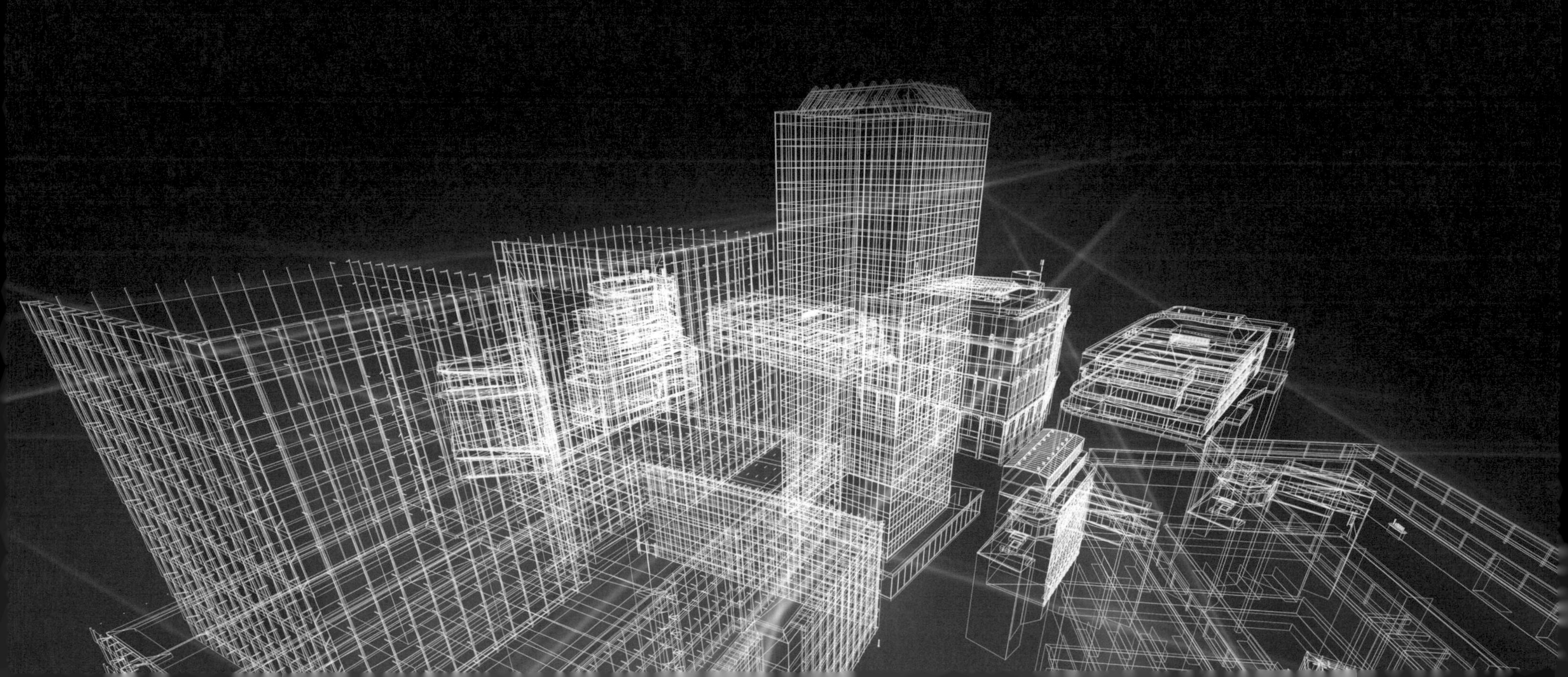

10.0 Mechanical/HVAC – All Areas

10.1.1 General

The process of acquiring a viable HVAC system shall involve decisions and actions taken during the predesign, design, construction and occupancy phases. The following describes some of the key aspects of HVAC system acquisition, in the order that would be typical for a new construction project:

- Establish HVAC related Owner requirements. The concept of a design cooling load shall derive from the need to determine an HVAC system size that, under extreme conditions will provide the specified conditions within the different spaces.
- The specialized design team shall carefully analyze system options, and guide the Owner to make the most appropriate solution to satisfy their business case.
- The following ASHRAE standards shall be referenced and complied with throughout the design development stage:
    - 90.1-2016: Energy Standard for Buildings Except Low-Rise Residential Buildings
    - 189.1: Standard for the Design of High-Performance Green Buildings
    - 62.1: Ventilation for Acceptable Indoor Air Quality
    - 55: Thermal Environmental Conditions for Human Occupancy
- The office building should provide a high degree of flexibility, controllability, and measurability. Systems should be able to efficiently turn down to support both localized and after- hours loads.
- Establish zoning requirements
- Make a preliminary system selection based on design issues/ intents/ criteria including code/standard compliance, and zoning requirements
- Calculate design cooling loads
- Select appropriate source equipment (to meet loads, intent and context)
- Select appropriate distribution approach (to meet intents and fit context).
- Coordinate HVAC components with other building systems.
- Rough-size equipment (fans, pumps, valves, air handlers, dampers etc.
- Conduct an energy analysis to optimize equipment selections and system assemblies.
- Final-size equipment based upon optimization studies.
- Consolidate final individual equipment elections into a cohesive whole.
- Develop appropriate control logic and strategies.
- Develop commissioning test protocols and checklists
- Provide benchmark performance data to the Owner.

In general, the final design should include, but not limited to:

- Good air quality
- Appropriate space acoustics
- Good thermal control
- Good solar/glare control
- High degree of ADA, DDA compatibility

10.1.2 Systems

The design of HVAC systems should start with a clear understanding of the outcomes required and/or expected by the Owner, and by the occupants. A comprehensive and cohesive statement of Owner's Project Requirements (OPR), in the form of a formal document shall be developed. The OPR shall address thermal comfort requirements for all occupied spaces, indoor air quality requirements for all occupied spaces, minimum acceptable energy performance, environmental preferences for constituent components, first cost, life-cycle costs, reliability, maintainability, building space and volume demands, aesthetics, and measurement and verification (M&V) requirements. Evaluate the benefits and limitations for each system or technology.

The systems designer shall take into consideration that most of the building will be leased to multiple tenants, analyze the pros and cons of cooling technologies for various building applications. The selected solution may include any combination, but not be limited to, the following:

- Centrally generated chilled water and distribution system, using vapor compression or absorption refrigeration cycles as a backbone of a cooling system.
- Chilled Beams for high sensible cooling environments
- Dedicated Outdoor Air Systems (DOAS)
- Variable Refrigerant Flow (VRF) Systems
- Under Floor Air Distribution (UFAD)/displacement ventilation systems
- Demand Controlled ventilation systems
- Car park ventilation (if applicable)
- Variable air volume supply and extract system for zone temperature control
- BMS monitoring and control systems
- Redundancy provisions for all central plant depending on the location, and subject to Owner approval.
- The office building shall provide flexibility, controllability, measurability, and feedback. lo accommodate the possibility of office tenants operating outside of the typical workday, the cooling systems shall be capable to efficiently turn down to support both localized and after-hours loads.

10.1.3 Design Criteria

- The design, installation and maintenance provisions should be in full compliance with all relevant International, European, and Local standards and regulations, and in particular any health, fire, safety, and environmental legislations. Emphasis shall be on building performance and adoption of energy- efficient building codes, such as ASHRAE Standard 90.1-2016.
- Designs are based on a host of variables beyond space type and understood loads including operating costs, first cost, lifecycle cost, and
- These systems should be designed and operated to achieve:
  - Design conditions indicated for each area as indicated in Table 10.1
- Design parameters should be achieved under conditions which do not exceed historical weather data for the location for more than 0.4% of the year.
- The life expectancy of the plant and equipment used should be in accordance with CIBSE, ASHRAE, BS 7542 or other recognized

International Standards, with an expected life in the order of 25 – 30 years.

- The buildings shall be designed to achieve an energy consumption level at least 30% below that attained by the ASHRAE Standard 90.1 – 2016 baseline building, if life-cycle cost-effective. If 30% reduction in energy consumption is not life-cycle cost-effective, the A/E shall evaluate alternate designs at successive decrements (for example, 25%, 20%, or lower) in order to identify the most energy-efficient design that is life-cycle cost-effective.
- Base external heat rejection equipment operates on external ambient design temperature at +5%.
- Operating costs should be minimized by utilizing passive building design, good insulation levels, efficient engineering plant and systems, appropriate zoning, effective control systems, selection of the most cost-effective fuels and heat recovery systems.
- Minimizing life cycle cost by utilizing natural energies (light/ solar) and building form, simplifying the services design and designs appropriate to the building location.
- Consider Environmental control of the large mall volume with ventilating and air-conditioning units at roof level pressurizing the mall space, from which retail shop units will draw their air which is then conditioned and recirculated within their zones.
- Exhaust air may be returned to high level from the retail floors by large ducts which climb up a shaft of the building floor plates, to discharge through the roof perimeter.
- Cooling unit configurations should be in a similar layout throughout the building, so that when maintenance personnel are familiar with one unit, they are automatically familiar with most of the equipment.
- All systems should have the capability, or at a minimum the ability to be modified for additional capacity in the future.
- The vertical and horizontal piping and ductwork shafts shall be sized for future systems. The stacking of vertical building shafts for exhaust and make-up air is critical to avoid costly extensions of shafts in the future.
- Consider the control of smoke in large spaces to be effected with the normal extract system, with recirculation shut down. A fire condition in the mall itself should send the main HVAC system into reverse, drawing smoke up to high level extract in the mall section.
- The systems proposed should comply with the following recognized International Industry Standards:
    - ASHRAE
    - CIBSE
    - NFPA (Smoke Control Systems)
- Expansion Capabilities
    - All major items of primary plant and service mains shall be sized to cater for the final scheme requirements of the Development Control Plan albeit plant may be installed on a

phased basis to suit the work stages. All primary equipment shall be designed with 25% spare capacity for future expansion

- All plant rooms and service routes within the buildings shall be provided with space for future expansion
- Where feasible pipe work systems shall be provided with valved and capped branches in suitable locations to allow extension of the system for future development without disrupting operation

The Designer shall be required to demonstrate and provide the acoustic values for all proposed systems to confirm compliance with and demonstrate that the relevant Owner standard has been complied with, and that the design noise criteria is not being exceeded. The design of the building and associated systems should be such that the energy consumption does not exceed the Source Energy Use Intensity (EUI) targets established by the relevant authorities.

Note 1:

(a) 0.35 air changes per hour, but not less than 7.5 L/s per person. For calculating the air changes per hour, the volume of the living spaces shall include all areas within the conditioned space. The ventilation is normally satisfied by infiltration and natural ventilation. Apartments with tight enclosures may require supplemental ventilation supply for fuel-burning appliances, including kitchen equipment and mechanically exhausted appliances. Occupant loading shall be based on the number of bedrooms as follows: first bedroom, two persons; each additional bedroom, one person. Where higher occupant loadings are known, they shall be used.

(b) Kitchens - 50 L/s intermittent or 12 L/s continuous.The air exhausted from kitchens may utilize air supplied through adjacent living areas to compensate for the air exhausted.

(c) Bathrooms - 25 L/s intermittent or 10 L/s continuous

The air exhausted from bath, and toilet rooms may utilize air supplied through adjacent living areas to compensate for the air exhausted.

The MEP designers should provide calculations to illustrate that these targets are being achieved.

10.1.4 Aims and Objectives

The aims and the objectives of the HVAC systems are to:

- Provide a comfortable odor free environment for occupants throughout occupancy period. Provide individual temperature control in each thermal zone, with adjacent spaces having cooling simultaneously available.
- To provide fresh air to all areas for the dilution of odors, and for occupancy needs. This should be pre-conditioned by Dedicated Outdoor Air Systems (DOAS).
- To remove vitiated air from the areas.

10.1.5 Thermal Zoning

Thermal zones are a key basis for HVAC system selection, and zoning decisions will play a critical role in occupant comfort responses. Thermal zones within the building shall be selected and should be provided with separate control.

The design team shall provide just sufficient zones to meet the owner's project requirements (with some flexibility for anticipated future needs). Fewer zones will result in some level of discomfort. More zones than necessary will increase the first cost of the project.

Thermal zoning should be established prior to the selection of a climate control system. The concept of thermal zones applies equally well to passive climate control systems as to active climate control systems. A zone may be a single room in the building, may consist of multiple rooms, or may be a portion of a room (a room may have more than one zone). The intent of zoning is to set up a control scenario that can respond to changing room loads and maintain thermally desirable conditions. Zone control in the HVAC system is most commonly initiated by a thermostat that senses room air temperature The Room/System Criteria Temperatures shall be as detailed in Table 20.1.3, and a minimum supply air temperature of 13°C (Cooling), with a maximum air velocity in the occupied space of 0.15 m/s shall be maintained.

The residential apartments should have a minimum of five thermal zones (based on solar orientation); would require to provide as many zones as there are apartment units.

In the case of an office, uniformity of ceiling heights, light fixture placement, grille locations, promotes flexibility in varying arrangements that can extend a building's usable life span. There are at least four types of offices that may need to be interchangeable within such generic space. The typical enclosed office has the privacy of four walls and a door. The bullpen office has repeated, identical workstations, with low dividers at about the height of the desk surface. The uniform open-plan office resembles the bullpen, but with higher divider partitions for added privacy. The free-form open-plan t has some individually designed workstations with divider partitions of varying heights (some spaces reflecting the varying status of workers). In bullpen and uniform open-plan office, the resultant uniformity is not always attractive to users, and diversity is often encouraged at a more personal level.

Diversity in the thermal conditions to be maintained, such as cooler offices and warmer circulation spaces, can be used to enhance the comfort of the office users.

Table 10.1 Mechanical Services Design Criteria

Area	Location	Space Conditions		Pressurization	Occupant Density #/100 m²	Fresh Air Quantity	NC Rating	Minimum Filtration Standard
		Deg. C DB	Max RH %					
Public Area	Entrance Lobby	24	60	5% Positive	-	-	40-45	F6
	Mall Common Areas	24	60	5% Positive	32.3	4.8 l/s/person	40-45	F6
	Retail Store	24	60	5% Positive	16.1	7.8 l/s/person	40	F6
	Barber/Beauty Salon	24	60	5% Positive	26.9	5.0 l/s/person	35	G5
	Prayer Room	24	60	-	100.0	2.8 l/s/person	40	-
	Public Washrooms	24	60	10% Negative	-	15 AC/hour	40	G5
Meeting & conference Areas	Entrance Lobby	23	60	5% Positive	-	-	40	F6
	Pre-function Areas	23	60	5% Positive	32.3	4.8 l/s/person	40-45	-
	Ballroom	23	60	5% Positive	71.4	11.0 l/s/m2	35	F6
	Auditorium Seating Area	24	60	5% Positive	161.0	2.7 l/s/person	25-30	F6
	Business Center	24	60	5% Positive	26.9	7.4 l/s/person	40	F6
	Exhibition Hall	23	-	5% Positive	43.0	5.3 l/s/person	40	F6
	Meeting Rooms	23	60	5% Positive	53.8	3.1 l/s/person	25-30	F6
Food & Beverage Areas	Restaurant Dining Areas	23	60	5% Positive	75.3	5.1 l/s/person	35-40	F6
	Kitchens	26	NC	Negative	-	10 AC/hour	45	-
	Fast Food Dining	23	60	5% Positive	108.0	4.7 l/s/person	40	F6
	Coffee Lounge	23	60	Neutral	108.0	4.7 l/s/person	35	F6

Table 10.1 Mechanical Services Design Criteria -*Continued*

Area	Location	Space Conditions		Pressurization	Occupant Density #/100 m²	Fresh Air Quantity	NC Rating	Minimum Filtration Standard
		Deg. C DB	Max RH %					
Fitness Facilities	Reception/ Lounge	23	60	5% Positive	-	0.8 l/s/m2	40	F6
	Gymnasium -Weight Room	23	60	5% Positive	See Table 4.2.1a 4.2.1b	4.21 l/s/m2	40	F6
	Studio - Aerobics	23	60	5% Positive		2.9 l/s/m2	30	F6
	Indoor Swimming Pool	24	60	10% Negative	-	15 l/s/m2 of wet area	40	-
	Changing/Locker Rooms	24	NC	10% Negative	-	10 AC/hour	40	F6
Office Building	Main Lobby	24	60	5% Positive	10.8	5.5 l/s/person	40	F6
	Reception Area	24	60	5% Positive	32.3	3.5 l/s/person	40	F6
	Open Office Space	24	60	5% Positive	5.4	8.5 l/s/person	35 - 40	F6
	Private Offices	23	60	5% Positive	-	4 AC/hour	30 - 35	F6
	Meeting Room	23	50	5% Positive	53.8	3. l/s/person	25 - 35	F6
	Technology Room	23	50	5% Positive	-	1 AC/hour	40	F8
	Washrooms	24	-	10% Negative	-	15 AC/hour	40	-
	Storage Space	28	-	-	-	4 AC/hour	40	G3
Residential Building	Apartment	24	-	Negative	-	Refer note 1	30 - 35	G3
	Bedroom	22	60	Negative	-	10 AC/hour	25 - 30	G3
	Living Room	24	60	Negative	-	1.6 l/s/m2	30 - 40	G3/F6
	Corridors	24	60	Negative	-	1.6 l/s/m2	40	F6

Table 10.1 Mechanical Services Design Criteria - *Continued*

Area	Location	Space Conditions		Pressurization	Occupant Density #/100 m²	Fresh Air Quantity	NC Rating	Minimum Filtration Standard
		Deg. C DB	Max RH %					
Back of House Areas	Administration Offices	24	60	Negative	-	8.5 l/s/person	40	F6
	Staff Dining	25	60	Positive	-	12 AC/hour	40	F6
	Staff Kitchen	27	60	Negative	-	10 AC/hour	45	G3
	Staff Changing	25	60	Positive	-	12 AC/hour	40	F6
	Laundry	26	60	Negative	-	20 AC/hour	40	F6
	Housekeeping	25	60	5% Positive	-	12 AC/hour	35	F6
	Maintenance Workshops	25	60	Negative	-	12 AC/hour	45 - 55	G3
	Plant Rooms	30 Max.	70	Negative	-	20 AC/hour	65	-
	General Storerooms	23	70	Neutral	-	4 AC/hour	45	F8
	Lift M/C Room	24	60	Negative	-	As specified by equipment supplier	45	-

10.1.6 Heat Recovery and Energy Conservation Objectives

The objectives of any heat recovery/energy conservation system shall be to reduce the running costs in a cost effective simple/fool proof way.The following is a summary of the minimum expected measures to be incorporated into the Mechanical Design. Control Requirements

- Automatic Control systems should be installed for all systems used to enable their effectiveness to be maximized and monitored.

Description

The following heat recovery and energy conservation systems shall be evaluated and should be provided unless there is an environmental and economic reason for not doing so:

- Where air handling units are not located adjacent to each other, consideration should be given to the use of heat reclaim coils and associated pipe work and dedicated pumps.
- Economy cycle operation in all AHU's to use fresh air for "free" cooling when outdoor conditions permit.
- Carbon Dioxide control of fresh air supplies through variable air volume supply, and extract systems of all public spaces.
- Two speed ventilation systems (or inverter driven fans) serving the kitchens for economy in "low" production periods.
- A proprietary kitchen exhaust system with injection nozzles, such as "Capture Jet", to reduce exhaust requirements by approximately 30% from conventional systems, and eliminate grease passing through the system utilizing high efficiency cyclone filters and "UV" filtration.
- Heat reclaim from chiller condenser for preheating of Domestic Hot Water System (DHWS).
- Resetting of chilled water supply temperature from the chiller when external conditions allow to maximize chiller efficiency and minimize latent cooling.
- DDC control systems via a central Building Management System (BMS) to optimize system control strategies.
- Power factor correction equipment shall be provided to the main electrical distribution control boards to automatically correct the power factor to unity, such that no tariff penalty is incurred under any load conditions. The equipment should be automatically staged in 50 kVAr units.

10.1.7 Source Components: Cooling

HVAC system selection shall take into consideration that the building is leased to multiple tenants, who pay for utilities, and the components of the Air-conditioning system shall focus on three conceptually (and physical different means of introducing cooling into the building.

A central chiller plant that provide a central source of chilled water generation, with a degree of redundancy, and operating on the vapor compression refrigeration or Absorption refrigeration cycle shall be the backbone of the Air Conditioning system. ASHRAE 62.1 shall be referenced throughout the design process.

The designer shall leverage rapidly developing and maturing technologies, edge device logic, and analytics to create a highly integrated and symbolic building that is predictive, responsive, and adaptive.

Description

The equipment should be sized to cater for:

- Fabric/solar heat gain - noting this diversification with the load
- Ventilation heat gain
- Infiltration heat gain
- Diversity of use and occupancy, allowing also for energy saving measures included within the design.

10.1.8 Control Requirements:

The principles of convergence and automation shall be two key principles of the building’s technology design. Data, voice, security, video, wireless systems, HVAC, lighting, electronic building controls, and audio-visual elements shall be built into core and shell or base building and designed to run on one infrastructure sharing ubiquitous networks that are robust, redundant, and secure. An internet protocol (IP) based IT network infrastructure shall be used for control system communications.

Wireless technologies such as sensors, thermostats, and equipment control should provide flexibility and adjustability features to the control systems.

The HVAC systems should utilize Direct Digital Control (DDC) type of controls that automatically monitor and control via a BMS, and their own proprietary control systems (where applicable) to maintain desired space thermal conditions, control and maximize efficiency of operation under all conditions.

All software application should reside and operate in the system controllers. The software applications shall be editable through an operator workstation, web browser interface, or engineering workstation.

The selected control system architecture shall allow components from several vendors operate over a BACnet- adapted Ethernet LAN (Local Area Network). It makes possible a system containing BACnet, LonWorks, Modbus and proprietary subsystems to work together. The wireless building control devices shall offer both protocol openness (ability to communicate) and application openness (ability to add new functions).

Neural networks that involve automation systems that are capable of learning from use, are to be considered. They will predict usage patterns, adjust operations in advance without needing specific commands from occupants. When the building use pattern is highly predictable, as with retail and commercial occupancies, these self- programming systems will learn very quickly how to anticipate needs while conserving energy.

In office buildings, the occupants shall be provided with more control and autonomy in determining when systems run and what temperature their space is.

Conference rooms and quiet areas should be controlled by occupancy sensors associated with the lighting system.

Humidity override should be provided in meeting rooms, board rooms, ballrooms etc., to override chilled water temperature re-set when necessary. Control should be via the BMS with no adjustments possible by the public. The function rooms should be provided with a variable air volume system to enable unused areas to be closed down, and to allow individual temperature control for each space, where thermal zoning is similar.

10.2.1 Building Management System (BMS)

Using automated sensors and controls, the building shall be linked to energy-consuming systems—such as lighting, HVAC, and humidity control —and managed holistically for greater efficiency and savings.

The BMS systems should primarily be composed of three tiers:

- Tier I: the primary bus, sometimes referred to as the management level is typically a wired solution and generally uses BACnet/IP protocol. At this level the operator interface with the system, and devices such as operator workstations, Web servers, and other supervisory devices should be networked together.
- Tier 2: the secondary bus and commonly referred to and generally uses BACnet/IP, BACnet/ MSTP, or Lon Talk protocols. At this level field controllers, programmable logic controllers, application-specific controllers, and major mechanical, electrical, plumbing, and lighting equipment, shall be connected.
- Tier 3: the field level, end-user devices such as thermostats and other sensors should reside.While wireless networks can be used at all levels, this level would be the most economical implementation for wireless technologies due to their ease or installation, flexibility, and ease of relocation.

The BMS should be capable to jointly monitor and control multiple systems by entering scheduling information once.

Operator Interface

A graphic interface that will allow an operator to view at a glance, the status of any building system, should be provided. The Operator Workstation or server shall conform to the BACnet Operator Workstation (B-OWS) or BACnet Advanced Workstation (B-AWS) device profile as specified in ASHRAE/ANSI 135 BACnet Annex L.

The web server or workstation and controllers shall communicate using BACnet protocol. Web server or workstation and control network backbone shall communicate using ISO 8802-3 (Ethernet) Data Link/ Physical layer protocol and BACnet/IP addressing as specified in ANSI/ASHRAE 135, BACnet Annex J.

The operator interface software shall be graphically based and shall include a minimum of one graphic per piece of equipment or occupied zone, and graphics that summarize conditions on each floor of each building. The graphic interface should Indicate thermal comfort on floor plan summary graphics using dynamic colors to represent zone temperature relative to zone set point.

10.2.2 Wired and Wireless Technologies

Major wireless technologies shall be used in conjunction with a most advanced Building Management System (BMS), reducing the cost of ownership and achieving a faster ROI. The BMS should be capable of being expanded to system-wide control of many devices across the distributed enterprise, using web-enabled applications to monitor, report, and manage energy use.

It is essential that the designer evaluates advantages and disadvantages of different choices for building automation—wired versus wireless, and various commercial wireless standards in use to optimize the control function to the needs and budgets of the Owner. The optimum solution for a building complex would be the use of a combination of wireless and wired technologies, using both EnOcean and ZigBee devices in a single network, and running wiring between buildings and widely separated zones. Zigbee should be based on the IEEE 802.15.4 standard.

The design, specification, and implementation of wireless networks should apply various countermeasures to maintain system performance, by mitigating interference from radio frequency (RF) devices that operate within the same Industrial, Scientific, and Medical (ISM) frequency band.

The selected BMS technology should provide provisions for data encryption, mutually authenticated communications, and robust key management to ensure that the wireless networks are inherently secure.

(a) **Securing the Network**

As with any network-based system, security exposures and converged system vulnerability shall be addressed. Web based support tools – similar to those used by IT professionals – that provide automatic – control system updates and security patches, among other support tools should be applied.

In the event that the facility systems, security systems, and business systems will share a common network, monitoring and network management are crucial factors that would ensure security and reliability of the facility.

Establish system redundancy by installing an additional server, will ensure that data and operations remain uninterrupted should a server malfunction.

(b) **Building Analytics**

A building analytics platform shall work alongside the BMS, to perpetually search and notify the significant abnormal or undesired patterns in data. The building analytics platform should include fault detection, fault categorization, and root cause determination. The system selected should include the following layers of activity:

- Identify conflicting systems that consume excess energy
- Forecast and monitor energy loads, and setup strategies to minimize demand charges
- Draw additional business-critical variables into the analytics tool for analysis – anything needed to measure or monitor from weather forecasts facility data and more
- Query large amounts of raw data and historical trend log data
- Identify specific equipment and systems that require servicing, repair or replacement
- Model the effects and ROI of potential renovations, repairs and retrofits
- Provide a predictive maintenance approach
- Predict, monitor and verify energy efficiency improvements
- Notify building system faults to identify and proactively resolve issues

10.3.1 Chilled Water Production/Distribution

The central chiller plant should comprise proprietary high efficiency air or water cooled, water chillers with additional coils for free cooling of the chilled/condenser circuit. The chillers should be complete with their own microprocessor control system to maintain the required chilled water flow temperature (including chilled water re-set), all necessary safety controls and be compatible with the BMS system. Refrigerant gases used should be in accordance with the Montreal Protocol, or current legislation if more stringent.

Chillers should be arranged for a turn down of 12½% or better. Chillers should be provided with variable speed drives for increased efficiency of operation and reduction in wear.

Chillers should be selected to achieve maximum efficiency levels following an in-depth study of the optimum chilled water temperatures, condensing conditions, and load profile to achieve an energy efficient design.

Cooling Towers should be proposed for rejecting heat, they should be so designed to prevent short circuiting of discharge and should be treated with twin biocides to prevent Legionnaires disease, and incorporate a continuous bleed of pond water. Where space permits, closed and/or hybrid circuit cooling towers may be considered as being an option.

In the chilled water circuit, should be incorporate a chilled water buffer vessel to limit compressor starting to a maximum of 6 starts per hour, where necessary.

Cold/freezer rooms for food storage should always be served separately, with individual chilling plant and cooling towers, unless efficient operation of the main chiller/condenser water system can be demonstrated.

Chilled water should be distributed at the highest optimum temperature to balance:

- Level of dehumidification
- Degree of purely sensible cooling
- Number and size of cooling units

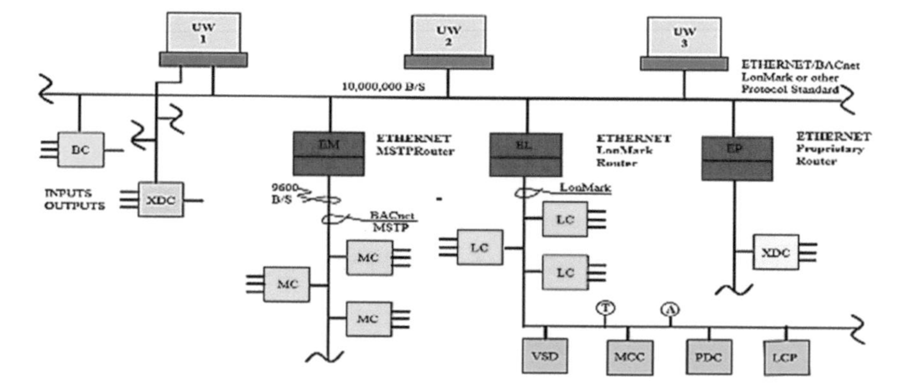

A	Intelligent Actuator	MCC	LonMark – compatible motor control centre
BN	BACnet native controller that connects directly on the Ethernet/BACnet bus	MSTP	Master/Slave Token Passing protocol
EL	Ethernet/BACnet – to - LonMark router	UW	User Workstation with Building Management and Control System (BMCS) and BACnet software and Ethernet card
EM	Ethernet/BACnet – to – BACnet Master/Servant Token Passing (MSTP) protocol router	PDC	LonMark – compatible power distribution centre
EP	Ethernet/BACnet to a proprietary protocol router	T	Intelligent temperature sensor
LC	LonMark - compatible controller	LCP	LonMark – compatible lighting control panel
VSD	LonMark – compatible variable speed drive	MC	BACNet/MSTP compatible controller
XDC	Proprietary digital controller		

Chiller control should include chilled water temperature reset based on ambient conditions for energy saving with humidity override if necessary.

The primary chilled water distribution should incorporate variable speed pumping.

The chilled water temperature set point at peak ambient conditions should be set as 7⅓C with a return temperature of 14⅓C. A chilled water bypass control valve should be provided to ensure the chillers minimum water flow rate is achieved always.

The chilled water should serve all main central air handling plant and fan coil units, if selected as the preferred terminal unit. The fan coil units should be selected to avoid latent cooling at the design conditions identified.

The configuration of the chilled water circuits should be such that clearly identified zones can be individually served, thus making it possible to manually shut down areas of the building for maintenance.

Modulating two port control valves should be installed on all FCU's and AHU's with a pressure operated by pass installed in the central plant room between the main flow and return, to maintain the minimum recommended flow rate through the chillers, and pumps controlled via a Variable Speed Drive (VSD). The DOAS air handling units may be provided with 3 port valves to assist in stabilizing the system operation.

Pipe work sizing and fitting selection should ensure that noise is not created within the system.

Heat rejection and air-cooled equipment should be selected to operate 5⅓C above the design external ambient conditions, and should reflect the actual location of plant, and the potential for local hotspots.

A satisfactory system of cleaning should be proposed for the condenser cooling water system, to prevent the build- up of corrosive elements and bacteriological growths, including positive measures to avoid Legionnaires disease bacteria. Adequate provision should be made for continuous bleeding of the system.

10.4.1 HVAC System Selection

The choice of HVAC system to serve the various leased tenant areas should depend on several factors including building configuration, ceiling void depths, plant room space, external space, utility charges and the extent of the work involved. A full evaluation of options should be undertaken to ensure the optimum system is selected for each subject.

The following systems are considered acceptable:

(a) All Air Systems (Preferred solution for air quality and energy use) with central air handling plant and recirculation where possible. These systems are most appropriate in areas that require high quantities of fresh air for occupants or equipment e.g. food court, meeting rooms, ballroom, restaurants and kitchens. Re-circulation should be incorporated where possible and variable fresh air provided, controlled via air quality (CO_2) sensors located in each area.

Air systems should be zoned utilizing Variable Air Volume (VAV) or similar (re-heat is not acceptable) per thermal zones previously identified, occupancy pattern and the operations carried out in that area.

Open office spaces that require large amount of ventilation should be provided with multiple variable air volume (VAV) boxes in the space.

Physical size of AHU's should be considered, and the practicality of maintaining large fans, motor and coils. Air systems should, where external conditions allow, have an evaporative cooler with suitable controls to enable free cooling to be provided utilizing outside air. These systems keep the major maintenance away from the occupied space, and hence avoid occupant disturbance and allow the use of free cooling when external conditions permit, resulting in reduced life cycle costs.

(b) If plant or ceiling space is limited (e.g. existing buildings), a Variable Refrigerant Flow (VRF) system should be considered, where one outdoor unit can be used with several evaporator units.

(c) Under Floor Air Distribution (UFAD) may be considered for of⅓ce buildings designed with raised ⅓oors, over traditional ceiling-based air distribution systems. The food court with special problems from heat, moisture, and aroma, as well as its schedule require a separate system.

(d) Two Pipe Fan Coil Units may be appropriate for areas that require low fresh air quantities e.g. offices, workshops, retail etc. and where the adjacent areas may require cooling at the same time. Two pipe vertical or horizontal fan coil units with individual means of isolation, filter, fan and condensate drain (preferably gravity fed) are an acceptable means of providing cooling.

Consideration should be given to providing common control valves serving several units in individual thermal zones, and will be BMS compatible. Units should operate on water-side control, with fan speed selected to meet the noise criteria shown in Table 10.1

(e) Fresh air supply to the thermal zone should be conditioned via a dedicated DOAS, to allow positive introduction of required ventilation air, which results in positive building pressurization. This pre-treated air should also be a method to supplement other devices such as active and passive-radiant and convective cooling technologies.

Some of the functions and methods of air introduction should include:

- Direct introduction of treated OA to each occupied space (OA direct to space)
- Introduction of treated OA to air-handling unit inlet (OA to AHU inlet)
- Introduction of treated OA to mechanical room plenum (OA to mechanical room plenum)
- Using treated OA for active chilled beams and other equipment needing primary air (OA to equipment).

(f) Commercial room thermostats shall have a touchscreen display for programming, scheduling, and monitoring. The thermostat shall have a built-in 365-day time clock with full-featured BACnet scheduling (SCHED-I-B). An adjustable delay on power up shall be available for soft start of systems on power loss and upon occupancy schedule changes. The ability to edit operating control parameters shall be protected via a user-definable security access code. Thermostats must incorporate non-volatile memory, so that in the event of power loss, all programmed operating parameters shall be unaffected without the use of battery backup. All control functions shall continue in the event of any/all communication failures.

Thermostats shall provide local communications in accordance with BACnet MS/TP ASHRAE 135. All BACnet objects and properties shall be published, open, and non- proprietary.

Indoor Swimming Pools: A HVAC system should be designed to provide a comfortable environment for users, to protect the building interior from the effects of chlorine/ condensation and to prevent chlorine smells spilling out to other areas of the building. The system should be designed to provide sufficient fresh air and exhaust as necessary to achieve this, and maintain the conditions specified in Table 20.1.3. Consideration should be given to providing a heat pump - de-humidifier unit, which de- humidifies the supply air, and uses the waste heat taken from the exhaust air (due to evaporation from the pool) for re-heating the air and heating the pool water. Other systems should only be provided following approval from the Owner. Kitchens: A dedicated AHU should be provided to serve the kitchen area. The unit should provide 100% fresh air (no-re-circulation permitted). This air should be tempered to room conditions (cooled) prior to supply to the kitchen. Mechanical extract ventilation should be provided to all kitchen areas. Stainless Steel Extract canopies should be located above cooking ranges to ensure that no spillage occurs. A minimum face velocity of 0.35 m/sec should be achieved across canopies. Suitable numbers of access doors should be provided throughout the extract ductwork to enable full cleaning to be carried out.

Make up air should be drawn from the dining areas or surrounding spaces, however should not be drawn over food services. If the kitchen has a significant heat gain, then further cooling may be required.

Kitchen hoods should be "Capture Jet" or equal with cyclone filter and Ultra Violet (UV) system to provide a 30% reduction in the exhaust rate from conventional systems, and eliminate grease passing through the system or equal approved system. The exhaust duct should be a dedicated system constructed from 16-gauge steel with drip proof weld joints, and clean out access doors are provided at all changes in direction, where sprinkler heads are located, and at every 6m maximum. Fire dampers are not permitted. The duct should be fire rated in accordance with the fire strategy and to meet NFPA 96 or local regulations, which are more stringent.

Dishwasher exhaust should be separate from the kitchen exhaust and should use stainless steel ductwork to prevent corrosion and graded to allow draining of condensation.

The exhaust systems should have variable speed, or 2 speed motors at a minimum to operate at low speed for energy savings at times of low use.

Administration & Back-of House - A/C and Ventilation: These areas should preferably be served by variable air volume all air systems from central plant, providing the zoning is similar. Each area should be provided with local automatic temperature control monitored by the BMS. Where zone requirements prevent all air systems from a central plant, each room / zone may generally be provided with a VRF system, or in ceiling two pipe fan coil units.

IT Rooms: See section 15.1.6

SECTION 11.0

Electrical Installation

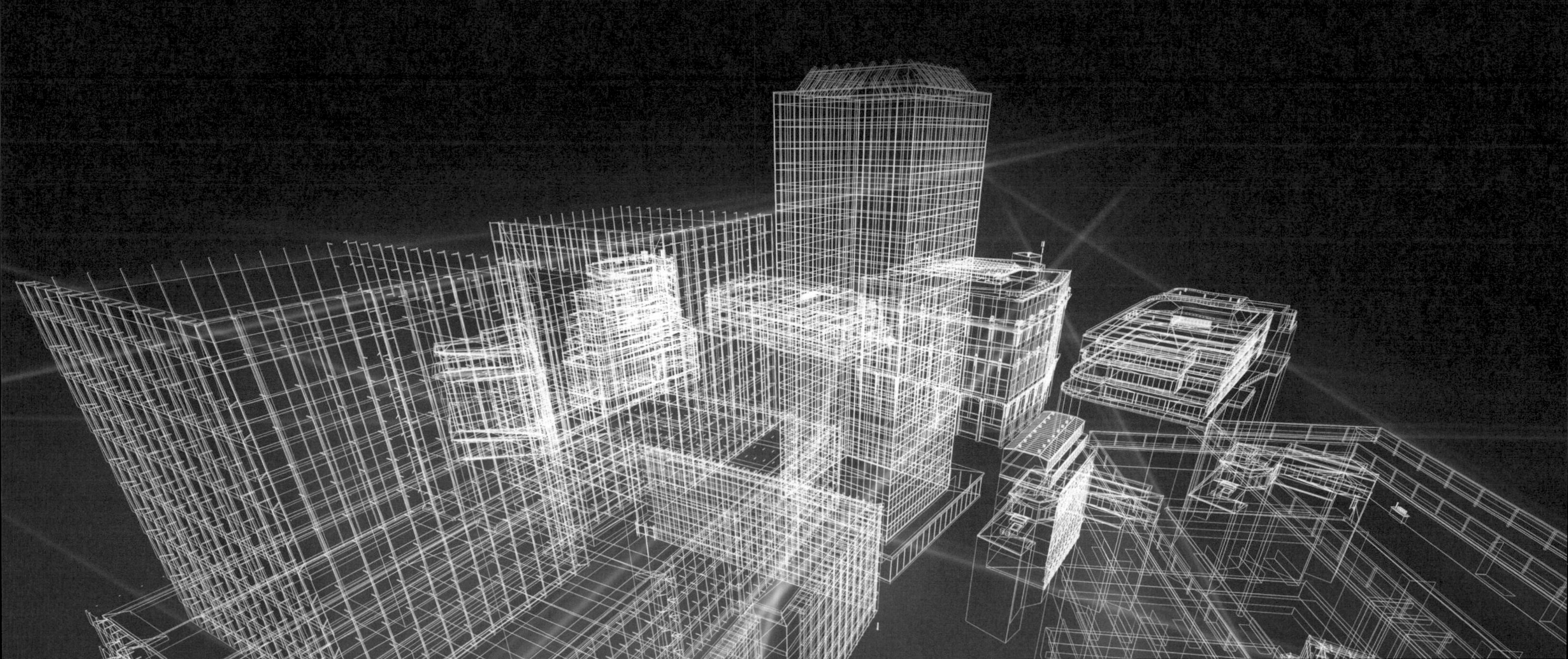

11.1.1 General

This section is a guide for Electrical Engineers and designers for the planning and design of the electrical power distribution, lighting and related systems, to provide a safe reliable power supply to maintain building operations.

11.1.2 Power/Electrical

The electrical service installations should be designed to serve all areas of the Facility, and provide for future expansion / flexibility. This should comprise 25% spare capacity within containment capacities, breaker space on all switch gear and individual transformer capacity (based on anticipated peak load requirement). For the purpose of this standard the following definitions are used:

- Extra - low voltage - Not exceeding 55V to earth.
- Low voltage - Above 55V to earth but not exceeding 1000V phase to phase.
- Medium voltage - Exceeding 1000V phase to phase but not exceeding 35kV phase to phase.

Where the electrical supply to the building is provided from the Electricity Supply Company (ESC) at Medium Voltage (MV) the substation arrangement shall be selected and installed in accordance with the supply company's requirements. The number rating and arrangement of the incoming supplies shall be based upon the following criteria:

- Availability and reliability of the local ESC's MV/LV network Maximum demand.
- Required redundancy factor.
- Availability and reliability of locally available equipment.
- Ability of local ESC to undertake their section of the works.

Installed redundancy shall be as illustrated by table 11.1.0

Table 11.1.0 Schedule of Ratings for Electricity Supply Types.

Supply Type	N° of Supplies	% Capacity Each	% Redundancy
MV	**1**	**100%**	**0%**
MV	**2**	**100%**	**50%**
MV	**3**	**50%**	**35%**
MV	**4**	**35%**	**25%**

Note: A single feeder is not acceptable.

Transformers

Transformer outputs should be brought out to a main switch board with bus couplers to enable segregation of the switch board for maintenance purposes. To reduce the material required to distribute power, while also reducing power losses, the MV primary distribution should be extended as close as possible to the load center. The reliability and redundancy should be increased by configuring multiple electrical feeds, taking into consideration the zoning within the mixed-use building. Branch circuit conductors that run from the final circuit breaker to the outlet or load shall be designed to allow a 3% maximum voltage.

Provide a transformer paralleling system to allow redundancy, so that the building shall be able to continue to operate without interruption should there be a loss of one transformer. This design follows the N+l policy for capacity. However, should there be a loss of two or more utility transformers, the building's standby generators should start and parallel, while the MV switchgear disconnects from the utility system. The building will then remain on the generator source until the utility source is restored, at which time the generators will parallel with the recovered source and, once the utility voltage has stabilized, reconnect the building load to it without interruption.

Energy efficiency should be a key component for transformers. The overall efficiency of the transformers shall be based on the selection of highest core-efficiency available.

An electrical load analysis should be carried out to detail the predicted loads for the building, by service and department area, based in W/m² floor loadings. This load analysis should include allowances for increase in electrical demand, and the sample table for completion by the designers is included in the Appendices to this section of the manual. Local code-mandated Watt-power densities with power for the equipment and lighting that will be installed and the sizing of the system components, should be reconciled.

If the load is known to produce harmonics, K-rated or harmonic mitigating transformers should be considered.

In addition to designing to relevant codes and standards, it is essential to identify unique requirements of the tenants as negotiated in the lease agreement.

Each area of the building should be provided with their own dedicated distribution panels. In general, systems should provide the following:

A reliable electrical supply: The electrical supply should have a reliability of - 99.98%. If a supply of this reliability is not available, then an additional electrical supply system should also be provided in the form of an independent electrical generating plant, or a totally independent power supply from an alternative generating source. An availability in excess of 99.98% is required (i.e. not more than one-hour interruption per year) to obviate the need for the generator.

The incoming supply, generator supply and electrical distribution should be arranged to prevent a total loss of power in the event of failure of a single item.

To ensure good quality power supply, the designer shall consider the effects of harmonics on electrical systems. Since electronic devices are nonlinear loads, they create both voltage and current distortions. The sustainable approach would be to implement a system-wide solution with harmonic mitigating transformers.

11.1.3 Electric Power Management system

An Electric Power Management System (EPMS) that will monitor the facility power distribution system for usage and quality, shall be provided. The EPMS should be a tool managing and ensuring the quality of the power, that is asource of power that is free from surges, sags, and outages that will affect the reliability and safety of the facility.

The EPMS should monitor the electrical distribution system, typically providing data on overall and specific power consumption, the quality of the power and event alarms. The EPMS should calculate usage trends, track and schedule maintenance, troubleshoot, and “bill back” metered power usage to specific tenants and specific users.

(a) **Monitoring**

The EPMS should monitor the power service entrance of the facility, switchgear, emergency generators, network protectors, switchboards, panel boards, Uninterrupted Power Supplies (UPS) etc.

The monitoring units should be microprocessor based, with on board memory, and can be programmed or have pre-set factors for monitoring, testing, and reporting sets. The monitoring of critical equipment such as switchboards and switchgear, should include digital metering and capability to monitor power quality. For power quality, the EPMS implementation should adhere to IEEE categories for power quality problems, including sags, swells, harmonics, interruptions, under and over voltages, and transients.

(b) **Central Operator Workstation**

A central operator workstation equipped with desktop computer and special application software shall be provided. The workstation should use data from system components to analyse and take action regarding the usage of power in the facility. The operator workstation should have at a minimum the following:

- Distribution System Graphics
- Real Time Reports
- Trend Reports

(c) **Historical Reports**

- Alarm Reporting
- Analysis of Electric Power Waveforms
- Determination and Initiation of Power Load-Shedding Strategies
- Communication with HVAC and Lighting Control Systems
- Usage Billing Software

(d) **Network**

Flexibility in network design and layout is critical. Physical topologies could be bus, star, and daisy chain configurations. Network components and devices should connect to the Ethernet network directly or through connectivity to a network router or gateway, essentially converting other network protocols, such as Modbus.

The network topology should result from criteria such as:

- The location of the components,
- The distances to be covered,
- The EMC requirements,
- Electrical isolation requirements,
- Conformance class, requirements,
- Requirements for increased availability and
- Consideration of network loads.

(e) **Transmission media**

Copper cables and optical fibres should be used for a wired connection of network nodes. The cable used must meet the requirements of the planned automation project.

When selecting the transmission medium, the designer should bear in mind an adaptation of the transmission medium to possible influences in the application area (e.g. chemical, electrical or mechanical).

In areas where electromagnetic interference may be present or significant earth ww w0potential differences are expected fiber optic (FO) connection should be used. Fiber optic connection can completely remove problems caused by electromagnetic interference (EMI) and/or ground equalization currents flowing in copper cable screens.

11.1.4 Energy Efficiency

ASHRAE 90.1 standard requirements shall be applied in the design for energy efficiency. The voltage drop shall not exceed 2% for feeders and 3% for branch circuits (Chapter 8.4.1).

- All the transformers shall meet the requirements for high efficiency transformers as outlined in ASHRAE 90.1 Table 8.1.
- balance the single-phase loads on 3-phase distribution systems. the unbalanced load should be designed to not exceed 2% unbalance.
- In applications where there is no intent to adjust the speed of the motor, a full load, or off-control scheme, a soft starter should be provided.
- In applications where motor speeds are varied between 50% and 100% to adjust for system demands, VFDs typically having an efficiency of 95% to 98%, depending on the type of VFD provided (6- or 18-pulse, active frontend, low harmonic, etc.)

11.1.5 Emergency standby power to dedicated supplies

Standby power systems designed to provide an alternate source of power if the normal source of power from the serving utility, should fail, shall be designed for reliability and in full compliance of NEC Article 701 or equivalent standards.

Additionally, the standby power supply system shall comply with the following Codes and Standards:

- IBC Chapter 27: Electrical
- NFPA 110: Emergency & Standby Power Systems
- NFPA 101: Life Safety Code
- NFPA 70: National Electrical Code
  - o Article 445: Generators
  - o Article 700, 701, & 702: Systems

The system should be capable of operating automatically or manually in an open transition or closed transition mode in conjunction with a load shedding system on the feeder breakers.

A separate custom control panel shall be provided to house the controls including a PLC, touchscreen and other control equipment. Hot Standby PLC equipment should be used, along with a control power transfer scheme to achieve high availability and a redundant system.

Standby backup power should be provided after a short break in electrical supply to serve essential, critical and other designated loads in the event of mains failure within 10 seconds.It may be necessary to divide the backup power system by building class. Low-rise office buildings may use only 90-minute battery backup power for life safety lighting, and fire alarm systems, and hydraulic or battery return for lifts. For mid-rise and high-rise buildings, a diesel or natural gas operated standby generator shall be provided. The designer should investigate the possibility of having a separation between generators used for the “base building” purposes and those generators used to backup tenant purposes.

The generator will be required to operate the loads indicated in the Table 11.1.1, with sufficient fuel storage for 48-hours continuous use. On-site fuel storage shall be designed in accordance with NEC 700.12(B)(3).

Note: Where power failures are historically extended, and fuel supplies are unreliable, storage capacity should be increased up to 7 days’ continuous full load operation.

Design

In view of the wide differences in applications, facilities, and conditions, the details of wiring and overcurrent protection of the electrical distribution system for on– site generation should be left to engineering judgement. There are however, some general guidelines to consider in the design:

- The design of the electrical distribution for emergency on– site generation systems should minimize interruptions due to internal problems such as overloads and faults. Subsets of this are providing for selective coordination of overcurrent protective devices and deciding on the number and location of the transfer switch equipment used in the system. To provide protection from internal power failures the transfer switch equipment should be located as close to the load utilization equipment as practical.
- Physical separation of the generator feeders from the normal wiring feeders to prevent possible simultaneous destruction as a result of a localized catastrophe such as a fire, flooding, or shear force.

Table 11.1.1

Description of Load	Essential Component (%)	Load Classification
UPS System	100	C
Life safety systems	100	E
Security systems	100	E
Fire pumps	100	E
Sump pumps	100	C
Sewage pumps	100	C
Smoke control/stair pressurization	100	E
Computer equipment not connected via UPS	100	O
Selected Lifts with any two simultaneously, including fire Lift	100	E
Domestic water pumps	100	C
Sewage Disposal plant	100	C
Data Centre air conditioning	100	C
Telephone system	100	C

Table 11.1.1 - Continued

Description of Load		Essential Component (%)	Load Classification
Lighting	Stair and Exit signs	100	E
	Switchgear rooms	100	E
	Emergency Generator rooms	100	E
	Plant Rooms	100	E
	Engineer's office	100	O
	Telephone operator's room	100	C
	Service areas	20	C
	Public areas	15	E
	Emergency Operations Centre	100	E
	Kitchens	50	E
	Data Centre	50	E
	Fittness Centre	50	E
	Residential Apartments	50	E
	Open PlanOffices	50	E

E = Essential C = Critical O = Other Designated Load

- Bypass–isolation transfer switch equipment so that transfer switches can be maintained or repaired without disruption of critical load equipment.
- Load–shed circuits or load priority systems in case of reduced generator capacity or loss of a single unit in paralleled systems.
- Fire protection of conductors and equipment for critical functions, such as fire pumps, lifts for fire department use, egress lighting for evacuation, smoke removal or pressurization fans, communication systems, etc.
- The security and accessibility of switchboards and panelboards with overcurrent devices, and transfer switch equipment in the on–site generator distribution system.
- Provisions for the connection of temporary generators (portable rental generator sets) for periods when the permanently–installed generator set is out of service or when extended normal power outages make it necessary to provide power for other loads (space air conditioning, etc.).
- The power transfer and control system should be interfaced with the building management system.

Synchronizing and Load Sharing Switchgear

To control the operation of the emergency power system, a fully automatic generator control and distribution switchboard should be provided.

Control circuitry shall be included in the switchgear to allow only "First Priority" loads (Emergency Loads) to be connected to the LV system as soon as first generator is connected to the emergency bus-bar. As the remaining generators are connected to the bus, this load control circuitry shall allow additional transfer switches to operate with time delays as described in the "Transfer Switch Schedule", and allow "Critical Loads" to be connected, whilst limiting the amount of load transferred to be within the capacity of the connected engine generators. In accordance with NFPA 110 requirements, any engine generator will start and assume load within seven seconds, and the balance of the generators are paralleled and will have divided load, within 10 seconds after the starting circuit is energized. In addition, as a fuel conservation feature, the control switchgear shall have the ability to start and stop the units in response to kilowatt load sensing, and to automatically shed and add load on a priority basis.

When local utility power returns, within proper limits, the control system shall bring the generators into sync with the utility power supply and close the utility Power circuit breakers. The engine generator sets shall then soft unload. The load shed/add feature incorporated in the generator control system shall activate if the load exceeds the engine generator on line capacity is exceeded.

Emergency Loads

This branch shall supply power to following loads per NFPA 70 and 99:

- Illumination of Means of Egress
- Illumination of Exit and directional signs
- Fire Alarm System
- Generator Set Accessories essential for generator operation
- Generator room lighting and power

- Switchgear rooms
- Lift machine rooms (selected lifts)
- Lift cabin lighting
- Fire and Sprinkler pumps
- Smoke control/stair pressurization systems

Critical Loads

- UPS
- Telephone system
- Walk-in fridges/freezers
- Central Plant room
- Engineer's Offices
- Security Office
- Sump Pumps
- Domestic water pumps
- Plant rooms – 50% lighting and 50% receptacles

11.1.6 Uninterruptible Power Supply (UPS)

Where sensitive load equipment is installed, a UPS shall be provided during the engine starting period to be a buffer between the generators and sensitive load equipment power. The operational goals shall remain the same regardless of UPS topology: The supply of uninterrupted power to sensitive, critical loads.

The uninterruptible power supply system should ensure no break in electrical supply occur to the following services:

- Tenant "IT" Systems (30 min)
- Telephone System (4 hours)
- BMS (24 hours)
- Security System (24 hours)
- Fire Alarm (24 hours)
- Emergency Lighting (3 hours)

The system should be sized to accommodate the electrical load for the equipment supplied plus a minimum of 25% spare capacity. The battery autonomy should be 20 minutes at full load.

11.1.7 Local Distribution/Protection

A distribution system should be provided to all items requiring a power supply. The distribution system should comprise main and sub-main distribution panels, trunking distribution systems, LV distribution cables, voltage regulators and power factor correction equipment.

The designer shall endeavour to develop flexible, sustainable electrical systems by maximizing resources. Specifically, the electrical systems design consumes resources both during construction and throughout the life of the building. Although a large part of the designs is driven by certain sections of the codes, these codes also contain some flexibility that allows designers to use fewer resources. ASHRAE Standard 90.1- 2016 could be considered as the most sustainability- driving standards.

The low voltage electrical switchgear system consisting of passive and active components should be designed with the following in focus:

- The passive components include steel frames, cover plates, barriers, and horizontal and vertical bus structures.
- The active components are critical and include power circuit breakers or fused devices, as these components are responsible for protection from overcurrent.

The low voltage switchboards should have a common grouping of "sealed type" circuit breakers in a common enclosure. The breakers should be directly connected to the bus and may be group mounted or individually mounted in their own compartment within the entire enclosure.

Switchgear shall consist of individually mounted and compartmentalized draw-out power, open-type circuit breakers. There should be physical barriers between the circuit breakers, and between the breaker and the bus.

Interfaces enabling connectivity to metering, breaker status, predictive information, control (on/off) shall be provided. Ethernet infrastructure communication system should be used with Modbus TCP/IP as access point for data and allow messages to be sent over the Internet. All low voltage electrical switchgear equipment should be connected to enterprise software packages (SCADA, EMS, BMS, etc.).

During the design development stage, the designer shall give particular attention to the following two active components and one passive component, as they have the most impact on maintenance and obsolescence issues.

- The steel / copper / or resin housing (passive, illustrated in blue)
- The circuit breakers, either removable or fixed (active, illustrated in orange)
- The protection relays (active, illustrated in red)

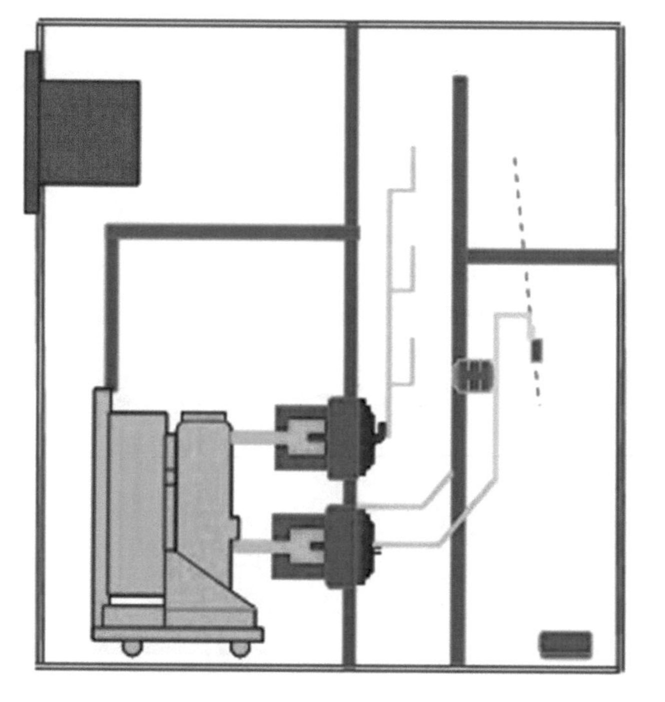

Figure 11.1.7

When designing the power distribution system, floor space appropriate for power distribution equipment while still meeting the NEC's requirement for safe working space in front, besides, and behind the electrical equipment, should be carefully planned. A few key NEC articles that address this issue are:

- Article 110.26 - Requires that sufficient access and working space shall be provided and maintained around all electrical equipment to permit ready and safe operation and maintenance of such equipment.
- Table 110.26(A) - Defines requirements for minimum depths for working space in low-voltage (0 - 600 V) installations.
- Table 110.34 (A) - Defines requirements for depths of working space for installations of 601 V and above.
- Article 110.26(A)(2) - Specifies that the width of the working space in front of the electrical equipment shall be the width of the equipment or 762 mm, whichever is greater. The width of the working space is a factor regarding worker safety. When the possibility exists to encounter live components, a worker must have adequate room to avoid contacting grounded components or incurring injury when retreating.

The horizontal through bus which connects the individual section bussing, and individual bus sections shall have ampacity ratings determined from UL 891 heat rise testing.

The power circuit breakers should be designed utilizing space-age materials packaged in compact fashion, with digital trip units. Digitization is essential to have enhanced equipment connectivity. This allows access to more detailed levels of energy management data, which enables better monitoring of energy consumption.

Facilities for electrical metering & monitoring shall be embedded within a circuit breaker.

Remote mimic panels shall be provided for circuit breaker control operations to minimize arc flash exposure of personnel.

A coordination analysis should be performed to provide power equipment with the required protection and minimize service interruption under overload or short circuit conditions. This analysis shall evaluate the electrical system’s protective devices including relays, fuses, circuit breakers, and the equipment to which they are applied.

Sub-main electrical distribution should be arranged to limit the failure of any single item of equipment or section within the system. This will be accomplished by the correct installation of adequate protection to achieve discrimination, together with segregation of equipment and cables as appropriate.

Power factor correction equipment should be included to correct the power factor such that no tariff penalty is incurred under any load conditions.

(a) Distribution Panelboards

The designer shall use the terms distribution or power panels for panels that feed other panels or large loads, lighting panels for lighting loads and lighting controls, and receptacle or branch panels for panels that feed socket outlets or smaller loads.

Harmonics

Harmonic distortions generally transmitted by nonlinear loads, such as Switch-mode power supplies (SMPS), variable speed motors and drives, photocopiers, personal computers, laser printers, fax machines, battery chargers, and UPSs should be corrected through a harmonic current mitigation strategy.

Since harmonic mitigation vary in complexity and cost and can be deployed individually or in combination, the strategy that makes the most sense for the facility should be implemented based on the loads it supports, budget, and the nature of the harmonic-related problems the facility will experience.

Table 11.1.2 Transfer Switch Schedule

Name	Branch	Priority	Time Delay settings (Seconds)			
			Engine Start	Transfer to Emergency	Retransfer to Normal Source	Engine Shutdown
ATS - 1	Emergency Loads	1	2	10	300	300
ATS - 2	Critical Loads	2	2	20	300	300
ATS - 3	All other Equipment	3	2	30	300	300

(b) **Metering Facilities**

All distribution boards serving loads in excess of 50KW should be provided with sub-metering.

11.1.8 Grounding and Bonding

Grounding and bonding should be provided throughout to prevent electrical ignition, minimize equipment damage, minimize electrical system exposure and to minimize business interruption, and ensure safe disconnection of the electrical supplies under earth fault condition. This should be in accordance with the 2014 National Electrical Code – Article 250 or equivalent British or European standards, where there is no conflict with local requirements.

The selection of a specific method of system grounding shall relate to various factors such as equipment cost, electrical system maintenance, safety of personnel, damage to equipment arising from ground fault conditions, system protective devices and their settings, and power continuity requirements of the facility.

11.1.9 Motor Control Center (MCC)

Each MCC shall provide a compact, modular grouping for motor control and electrical distribution components. A MCC should be implemented where a central control point is needed to remotely operate multiple loads. Distributed control systems (DCS) or programmable logic controllers (PLCs) should be used to provide this control and data acquisition functionality.

Intelligent Motor Control Center (IMCC) having Intelligent devices imbedded in the MCC shall provide network communications and functionality that is not available on standard devices, such as network configuration, diagnostics, process information, and advanced protection for each unit. Networks with high bandwidths should be used to facilitate the transfer of large amount of information available. The IMCC supplier should be able to integrate with multiple networks.

Motor control equipment shall be designed to meet the provisions of the National Electrical Code (NEC), and code sections applying to industrial control devices Article 430 on motors shall be fully applied. Standards established bythe National Electrical Manufacturers Association (NEMA) shall assist the designer in the proper selection of control equipment.

11.2.1 Lightning Protection

A lightning protection system should be provided for protection of the occupants, structure and its contents in the event of a lightning strike. This should be in accordance with 2014 NFPA 780 or equivalent British or European standards, where there is no conflict with local requirements. The design engineer should use the Faraday cage or Franklyn Rod type lightning protection systems. Where standards permit, the steel frame or reinforcement of a building may be used as the down conductor system. However, the design engineer should consider the requirement for exothermic welded bonds between steelwork sections to ensure a reliable earth path.

11.3.1 Lighting

(a) **General**

In many instances, energy codes require lighting engineers to design lighting systems that meet prescribed power allowances, use daylighting controls, control spaces through occupancy, and specify and perform functional testing in their lighting designs. ASHRAE and the International Energy Conservation Code (IECC) both require spaces surrounded by ceiling height partitions (walls) to have an individual manual control (switches/ dimmers). The control must be within the space or remote located with an indicator that identifies the space/ area it serves. Exemptions to this requirement are for areas that must be continuously illuminated for safety/security and corridors or stairways used for means of egress.

The following design criteria are to be considered as minimum requirements which may be exceeded by local practices. The design engineer shall liaise closely with the Architect, Interior Designer & Specialist Lighting Consultants with regards to the selection and location of decorative luminaries and effects, both within and outside the building. The specialist lighting designer shall influence some of the building's glazing properties, shade controls and other features to improve the benefits of daylight harvesting.

(b) **Control Systems**

Lighting systems shall have control steps between 30% and 70%, which could be accomplished with a number of variations such as switching alternating lamp, dimming ballast/driver, or stepped ballast/driver with even illumination in the space. Lights in corridors, electrical/mechanical rooms, public lobbies, washrooms, stairways, and storage rooms shall be exempted from this requirement.

A scalable architectural lighting control system should be designed for the control of architectural lighting for use in the various architectural environments, from simple meeting rooms, retail spaces to networked systems in large venues such as the Conference and Exhibition Centre. The hardware shall include modular dimming and processing panels, options for programmable control stations, and versatile interface devices.

In public sitting and transition areas such as an Atrium lobby, dimming controls should be used to make these spaces unique and inviting. In the ballroom and specialty restaurant spaces, dimming should be used for mood setting. In meeting rooms and conference rooms, dimming should provide the flexibility needed to accommodate a variety of presentation media.

In retail spaces the lighting specialist shall integrate all control features associated with lighting via a centralized intelligent control panel or a distributed system. Specific inclusions in such systems should include occupancy detection, dimming, daylight linking and scene-setting. Scene Controls should be used to operate several device series and set multiple levels or actions via one command and typically used to:

- Create mood lighting and quickly transform in-mall ambience.
- Schedule-switch or dim several lighting areas at a day's start or end.
- Enable/disable occupancy sensors when moving from working to nonworking hours' control.

All public area lighting requirements should have lighting control platforms that can be activated via a combination of manual switches and/or automatic controls via time scheduling, light level or occupancy sensors.

The lighting control technologies, including those that integrate occupancy recognition shall be tied into the property building management system or a micro- processor based centralized programmable lighting control. The designer shall endeavor to apply open protocol controls and avoid the limitations of proprietary control systems. In the spaces where occupancy recognition is applied, dual technology sensors and controls (Infrared and ultrasound) shall be applied to eliminate false occupancy sensing and ensuring proper systems operation and function. When an automatic time switch control device is used, an override should be located in a readily accessible space with a maximum override up to 2 hours.

Occupancy sensors are required in all conference/ meeting rooms, employee lunch and break rooms, private offices, storage rooms, janitorial closets, and other spaces $28m^2$ or smaller enclosed by ceiling height partitions.

Lighting circuits should be monitored via a central location to ensure everything is functioning correctly and that circuits that fail can immediately be identified.

(c) **Daylight Harvesting**

Daylighting shall be effectively integrated with the electric lighting system. Daylighting control and automation systems shall sense the amount of natural daylighting and dim or shut off artificial lighting in the day-lit space, and provide the designed amount of work area illumination. The designer shall identify lighting dimming technologies that have built-in provisions for daylight control schemes

In spaces with plenty of natural light, daylight harvesting shall use continually monitoring and adjusting artificial and natural lighting to enhance occupants' comfort by eliminating variations in lighting levels.

A total light management system that includes automated shade control should form the lighting design strategy for addressing both energy and glare management. The best solution should combine automated shade control with solar-adaptive software, and cloudy-day/shadow sensors that allow the shading software to evaluate and respond to real-time daylight conditions. It is important to choose a shade manufacturer that offersa broad selection of fabrics, allowing the lighting designer/architect to choose the appropriate transmittance level based on the building's location and orientation.

Light level sensors should be used to determine lighting levels according to the sunlight filtering through windows in specific areas. Should a dimmer detect less than optimal artificial brightness in an area, it brightens lights to the desired brightness value. Similarly, when natural brightness begins to fade outdoors, the lighting control system should adjust the artificial light in small steps. Should they wish to, users can override these automatic light controls by means of a simple push-button.

(d) **Exterior Lighting**

Exterior and feature lighting at entrances, ornamental gardens and lobbies shall be provided through seamless integration and lighting transitions via a DMX interface gateway which should be compatible with different manufacturers' LED lights and theatrical equipment.

Building grounds lighting luminaires over 100 watts should have lamp efficacy of at least 60 lumens/Watt.

The exterior light fixtures should be designed to automatically switch-off based on daylight, and any decorative façade and landscape lighting must be automatically shut off between midnight and 6 am, with the exception of security lighting. The exterior lighting should also be controlled by a combination of photo sensor and astronomical time switch. The basic components of the exterior control system should seamlessly interface with the control strategy applied on the interior of the building for a total building lighting control solution.

The lighting control systems shall be tied to the building management system, with astronomical time switches, and photocells providing an input to the system. Dusk to dawn operation shall be accomplished with a photocell controlling all of the fixtures, or lighting fixtures that have integral photocells or a combination of the two approaches.

External lighting should be provided for all pedestrian and vehicular access routes to the building, service and car parking areas. Road lighting should be in accordance with local codes of practice.

Decorative lighting and/or flood lighting should be provided to the major elevations of the building, and to landscaped areas and signage. All external lighting should be time switch and photo-cell controlled, with manual override switches being provided in the Security Office. Separate controls should be provided for the following:

- Road, car parking a nd delivery area lighting
- Pedestrian area lighting and building entrances
- Decorative and flood lighting
- Illuminated signs and displays
- Water features/ artwork / sculpture
- Roof and external plant areas

Control of parking garage fixtures should have automatic shut-off in parking garages based on both occupancy and exposure to natural daylight. The occupant control should be by one or more devices that automatically reduce power of each fixture by a minimum of 30% when no activity is detected within a zone. The light fixtures shall be equipped with onboard occupant sensors. The onboard sensor should signal the fixture to reduce the light output to a preset level. The lighting designer should ensure to not jeopardize the safety of the garage occupants. Guidance notes described in the latest version of ASHRAE 90.1 shall be applied, so that the controls are triggered far enough in advance so that a car or pedestrian is not entering a dark area before the fixtures are triggered to react. Additional controls are required to automatically reduce lighting levels of fixtures located within 6m of a perimeter opening exposed to daylight. Similar to the indoor application of daylight control, parking garage daylight sensors should be installed to reduce the light output in response to daylight.

(e) **Luminaires**

Taking into consideration the Owner's drive for energy efficiency, the design engineer shall advise the design team to maximize the use of energy saving long life luminaries. Solid-state lighting technology, such as Integrated LED lighting solutions that establish consistent color temperature in spaces, shall be implemented.

LEDs selected should have the capability for continuous dimming from 20% to 100%. However, linear fluorescents are required to have a minimum of four different illumination levels (20% to 40%, 50% to 70%, 80% to 85%, and 100%) for lamps greater than 13 W.

LED Recessed Troffers 3500 to 4000K lamps with electronic ballasts should be considered for commercial office interiors. All digital ballasts shall comply with the Digital Addressable Lighting Interface (DALI) protocol. In general office areas, any one of the following two basic approaches may be applied:

- Uniform layout
- Task-ambient design

In private offices, a task-ambient design approach is required.

In an environment such as retail store where color is important, the designer shall apply a high Color Rendering Index (CRI) to display the true colors of the displayed item.

LED luminaires shall be tested and certified to IMS LM-79 and LM-80 standards. LM-79 standard guidelines, and shall certify LED luminaires for light output, energy use, and color spectrum. LED luminaires shall have efficacies between 70 and 90 lumens/W. The minimum life of an LED luminaire shall be at least 70% lumen maintenance (i.e. L70) at 50,000 hours of operation, with a light loss factor of not more than 0.7.

Video Conferencing Spaces: Solid-State-Lighting shall be installed in rooms used for Video Teleconferencing. The lighting shall accommodate the visual requirements of the people and the camera, and provide visibility of any display material to remote participants.

Strategically placed LED fixtures with appropriate control solutions, such as dimming controls should be installed to produce required vertical illumination, create uniform lighting on surfaces and participants. The control solution shall have the ability to adjust contrast in the space. The IES (Illuminating Engineering Society) standard IESNA DG – 17 -05 shall be used for guidance.

Image clarity shall be enhanced by implementing a solution that combines fluorescent fixtures with dimming ballasts and/or LED fixtures with dimmable drivers in various form factors to fit different space types and applications

Table 11.3.1 Lighting Power Densities (LPD) Using the Building Area Method

Building Area Type	LPD watts/m2
Convention center	11.63
Performing arts theater	14.96
Dining: bar lounge/leisure	10.66
Dining: cafeteria/fast food	9.69
Dining: family	9.58
Health & Fitness Club	10.76
Residential Apartments	6.46
Offices	9.69
Retail	15.07
Maintenance Workshops	12.92
Store Rooms	7.10

Table 11.3.2 Schedule of Lighting Levels

Area	Location	Intensity Lux	Working Plane
Shopping Mall	Main concourse	300	Floor Level
	Entry/vestibule	300	Floor Level
	Food court	300	Table Top
	Entertainment areas	500	Table Top
	Washrooms	100	Floor Level
	Vertical transportation	300	Floor Level
	Side arcade	300	Floor Level
	Kiosks	1000	Table Top
	Barber shops and beauty parlors	1000	Floor Level
Auditorium	Assembly	200	Floor Level
	Social Activity	100	Floor Level
Exhibition Halls	Assembly	200	Floor Level
Conference Rooms	Conferring	500	Table Top
	Critical Seeing	1000	Table Top
Fitness Centre	Gymnasium	500	Floor Level
	Excercise Studios	1000	Floor Level
	Locker Rooms	200	Floor Level
	Indoor Swimming Pool	1000	On Pool Deck
Office Block	Lobbies, Lounges and Reception Areas	200	Floor Level
	General and Private Offices	300	Floor Level
	Conference Areas	500	Floor Level
	Circulation Areas	200	Floor Level
	Video Conferencing	500	Table Top

Table 11.3.2 continued

Area	Location	Intensity Lux	Working Plane
Residential	General Lighting	175	Floor Level
	Dining Room	430	Table Top
	Kitchen	650	Table Top
	Bathroom	650	At the Mirror
	Home Office	550	Table Top
Back of House	Stairways	200	Floor Level
	Service Corridors	1000	Floor Level
	Freight Elevators	200	Floor Level
	Storage Rooms	200	Floor Level
	Plant Rooms	500	Floor Level
	Main Control Room	500	Table Top
	Security Control Room	750	Table Top
	Commercial Laundry	500	Floor Level
	Maintenance Workshops	500	Floor Level
Outdoor Facilities	Building Entrances	100	Floor Level
	Gas/Fuel Storage Yard	100	Floor Level
	Gardens	50	Floor Level
	Outdoor Swimming Pools	150	On Pool Deck
	Loading Bay	200	Floor Level
	Outdoor Parking Areas	5	Floor Level
	Indoor Parking Areas -Entrance	500	Floor Level
	Indoor Parking Areas - Ramps	20	Floor Level
	Restaurants and Dining Areas	50	Table Top

11.4.1 Security Systems

General

There is no single formula for the design of defensible space. A thorough understanding of the physical and social environment of a neighborhood is required: What is the physical layout? Who is coming and going? Who belongs and who doesn't? What are the dynamics of the problem? What are the neighborhood routines? The Owner's Security Policy shall be the foundation for designing a security system.

In general, three aspects of defensible space design shall be applied: territory, access and surveillance. Territory refers to private and public space. Territory should be established by drawing distinctions between spaces.

Access shall refer to providing and restricting access; that is control. Surveillance refers to seeing and being seen.

The Owner shall engage the services of a security expert and design professional with particular security experience in protecting public facilities.

Unlike traditional offices, mixed-use buildings typically feature a combination of retail, offices, and dining spaces on the ground level with living spaces – generally apartments above. Without the right security solutions in place, it would be difficult to make both customers and guests feel welcome and maintain the safety and security residents and staff.

Although access control systems can balance these concerns, it is important to be aware how those systems can impede authorized access for the elderly and people with disabilities. It is important for the design team to be aware of the challenges these systems present and incorporate appropriate design solutions.

It is critical to include the relevant experts in the planning process. Typically, the door hardware consultant, the security consultant, the integrator, and the electrical engineer should be involved. Accessibility should be emphasized as a guiding principle planning process.

11.4.2 Integrated security systems

The specialist security systems consultant shall design and IP based integrated security system and enterprise integrated security system [which is also IP-based. The security systems should have the following three major defining attributes:

- Integrated security systems should comprise numerous subsystems to get the into one complete, highly coordinated, high-functioning system. Typical subsystems should include an alarm, access control, closed-circuit video, two-way voice communication, parking control, and other related systems.
- System integration should involve bold the integration of components and the integration of functions. High-level functions should be obtained

by integrating components into a comprehensive working system, instead off individual disconnected subsystems.

- Design convergence-based integrated security systems by integrating security systems that utilize TCP/IP Internet infrastructure as the basic communications media.

Each element of the electronic security systems should act as a deterrent, perform detection, assist in assessment of the event with regard to severity, assist in reaction, and gather evidence.

Alarm/Access Control Systems

Access control shall be a service that provides the following primary functions:

- Controls entry to /exit from an area of the facility,
- Tracks and logs personnel throughout a facility
- Eliminates key and lock cylinder replacement costs when employees are hired or Terminated.

Before examining problems and solutions relevant to access control systems, it is essential to identify the type of openings they may be applicable. Openings common in mixed-use facilities may include, but not be limited to:

- Main entrances, where a single credential provides access to multiple openings;
- Resident entrances, which require strong and secure locks joined with stylish interior; door hardware
- Stairwell and emergency exits, which need durable door hardware able to perform with minimal maintenance;
- Common areas which must balance ease of use and proper egress compliance;
- Conference centers and business offices, which typically remain unlocked and accessible during business hours while the rest of the building remains secure;
- Areas with unique requirements such as swimming pools, fitness facilities, and parking garages;

Access control products must comply with same code requirements as mechanical hardware, and the design team should consider whether potential products will be appropriate for occupants of all ages and abilities.

These systems should be designed to restrict access into protected property, but at the same time, not interfere with the safety requirements of the building. Electronic access control equipment should be designed to comply with the latest edition of NFPA 731, and all control equipment shall be listed in accordance with UL 294.

Identification devices shall include card/key/barcode/ radio frequency identification readers, keypads and biometric readers. The designer shall select the proper technology for the application to protect the premises.

For optimum user ability, the designer should be aware of reader placement and locking systems.

Control of egress should comply with the requirements of the applicable codes and standards based on the occupancy classification and based on usage of the property. Where delayed egress is utilized in conjunction with an electronic access control system, equipment shall be listed for the purpose and installed in compliance with the applicable codes.

The facility would require creating Multi-Tenant applications with an environment of independent control for a number of different tenants using a single system. This is necessary to manage and report on events from the system. Each tenant should manage and control their own facility, while the property manager has control over common doors that all tenants utilize. Each tenant should manage their own personnel by using the Access Control Database Program on one of their own computers.

Delayed egress doors should be installed in retail areas. The delayed egress doors must unlock during a fire alarm, or when the control panel is in trouble mode.

Open architecture electronic locking systems which are designed to easily accept additions, upgrades and replacement of components, should be selected. These systems must provide multiple, interchangeable credential reader modules, as well as interchangeable off-line networking modules – both wired and wireless,

11.4.3 CCTV and Digital Video Systems

As part of the larger facility security plan, video surveillance systems shall be provided. Closed Circuit Television (CCTV) and analog/digital imaging systems shall be designed to provide positive visual identification of a person, object or scene. IP surveillance is the most flexible and future-proof option for the security and surveillance installation.

(a) Network Cameras

The designer should select cameras that meet the needs of the facility and installation. This includes cameras that can be pan/tilt/zoom, vandal-proof, weather-resistant, or fixed-dome products. Each type of camera should be capable of being blended into the IP Surveillance system, to create a total package that solves your security needs. Power over Ethernet (PoE) may be made available for fixed installation IP cameras. While power to some camera deployments will be an essential requirement (Pan-Tilt-Zoom housings, wireless cameras and cameras that require fiber connectivity due to distance).

The designer should ensure to select a network camera that has open interfaces (an API or Application Programming Interface), which enables a variety of software vendors to write programs for the cameras.

Advanced network cameras that have built-in motion detection and event handling should be provided where intelligent visual surveillance systems are required to provide an automatic interpretation of scenes, and to understand and predict the actions and interactions of the observed objects based on the information acquired by the cameras. In addition, more intelligent algorithms - such as number (license) plate recognition, people counting -- are to be integrated into security and surveillance systems.

The number of cameras may vary greatly depending on the coverage requirements and the nature of the usage of the facility.

(b) **Compression**

It is vital to select the right compression for the video signal, which includes choices between proprietary or industry standard modes such as Motion JPEG or MPEG-4.

Video management tools that will take into consideration available bandwidth, storage capabilities, scalability, frame-rate control and integration capabilities, should be employed.

(c) **Video Management**

The video management system is a very important component of IP surveillance systems because it effectively manages video for live monitoring and recording. Video management requirements will differ depending on the number of cameras, performance requirements, platform preferences, scalability, and ability to integrate with other systems. The designer shall select a solution that typically range from single PC systems to advanced client/ server-based software that provides support for multiple simultaneous users and large number of cameras.

The video surveillance system should be capable of being integrated with access control devices. The access control system should allow video to be captured at all entrance and exit points and for pictures in a badge system to be matched against images of the person actually using the access card.

The designer shall select a network video system that allow for open systems with video management software that can be installed on a PC server platform.

The system should be fully scalable, where cameras can be added one at a time, and there is no limit to the number that can be added or managed.

The video management system should use a Web interface to access the video from any type of computer platform, using the proper safeguards such as password protection and IP address filtering.

Video management software should support network cameras from multiple vendors to ensure flexibility.

(d) **Network**

The network design should be specific to the needs of the facility and the specific installation. Beyond the actual cameras, it is essential to consider IP addressing and transport protocols such as TCP/IP along with transmission methods, bandwidth, scalability and network security.

A highly available network is a network that provides connectivity at all times. As applications, have become more critical, the network has become significantly more important to businesses. A network design should provide a level of redundancy where no points of failure exist in critical hardware components. This design should be achieved by deploying redundant hardware (processors, line cards, and links) and by allowing hardware to be swapped without interrupting the operation of devices.

Wireless networking should be a preferred option within the facility where there is a need to move cameras to new locations on a regular basis. The technology should also be used to bridge buildings without expensive ground cabling, or to add cameras in difficult to reach locations. Wireless local area networks (WLANs) should be the basis for wireless networks.

In general, the designer should avoid loading a network to more than 50% capacity, and avoid risk of overloading the network. When building a new network, or adding capacity to an existing network, the designer should build in 30% to 40% more capacity than calculated. This will provide flexibility for increasing use in the future.

(e) **Storage**

The final stages in a surveillance system these storage and retrieval. Considerations when determining storage requirements should include frame rate, the amount of time the video needs to be stored, the required redundancy, and which type of storage that fits best, e.g. a storage area network (SAN), or network attached storage (NAS).

Video surveillance shall be optimized through technology that consists of a collection of dynamic features that analyzes and optimizes the network camera's video stream in real time. Scenes containing interesting details should be recorded in full image quality and resolution while other areas are filtered out, to optimally use available bandwidth and storage. Important forensic details like faces, tattoos or license plates should be isolated and preserved, while irrelevant areas such as white walls, lawns and vegetation are sacrificed by smoothing in order to achieve the better storage savings.

The technology should automatically adapt to PTZ camera movements. Bandwidth peaks should be avoided using a dynamic rate controller that will be automatically enabled when the camera is being panned, tilted or zoomed.

The designer shall allow redundancy in a storage system for video, or any other data, to be saved simultaneously in more than one location. This should provide a backup for recovering video if a portion of the storage system becomes unreadable.

(f) **System Integration**

The security systems shall be fully integrated to a unified system. The unified system shall have a single platform that manages a variety of security devices, including cameras, door controllers and readers, intercom devices, intrusion panels and all other devices.

Dashboards which are software based, web-based or cloud-based, shall be provided for real time reporting, and managing the security systems in a holistic manner. These dashboards shall provide advanced analytics on processes and services such as how effective the security systems have been performing, downtime, system exceptions, hardware or processes experiencing excessive service calls and other information, similar to a report card on system status.

The dashboard shall additionally monitor operations to continually assess alarm conditions and any ongoing issues in monitoring operations, pinpoint problem areas or consolidate alarms across the facility.

(g) **Security**

Securing video is one of the most important steps in creating a successful IP surveillance installation. Since all security and surveillance applications contain sensitive information that should not be available to anyone with an Internet connection.

Understanding and choosing the right security options - such as firewalls, virtual private networks (VPNs) and password protection – is vital to eliminate concerns that an IP surveillance system is open to the public

Wireless networks shall be secured by using the most appropriate solution for the facility.

In addition to protecting data, it is critical to control access to the system via a Web interface or an application housed on a PC server.

11.5.1 Energy and Water Metering

Energy metering facilities should be provided to allow the energy and thermal performance of the facility to be recorded and to establish baselines for performance and management, energy savings verification and air conditioning plant optimisation and control. An Enterprise Energy Management System (EEMS) shall be applied to consolidate all energy related data (sources, cost, control and monitoring points), and provide tools to access and interact with the data.

Cellular wireless wide-area network (WWAN) technology shall be applied where it is identified to provide distinct business and technical advantages over conventional radio communication networks. By enabling two-way communications between meters and utilities provider and providing accurate information about the status of the utility supply system, smart meters play an important role in energy conservation.

“Smart”meters shall use Advanced Metering Infrastructure (AMI) to allow for two-way remote communications with meters, using either point-to-point cellular WWAN technology deployed at the meter, or “fixed-network” solutions with cellular gateways connecting groups of meters using low-power

RF mesh networks. Smart metering shall typically include meter data management (MDM) software and services that provide utilities with end-to-end capabilities to remotely read meters and integrate data with back-end systems.

Wireless communication networks shall be configured for network security issues related to data integrity and information security when connected to standard data networks. To achieve these network security requirements, the communication protocols should include appropriate levels of information assurance and security, through control protocols that have been extended with a set of network layer security messages that provide data confidentiality and integrity, device authentication, data hiding, and user authentication.

Central energy plants that meter both chilled-water and heated-water circulation systems, energy shall be determined as the product of the fluid's mass flow and the corresponding temperature differential through the system being metered. Positive displacement, differential pressure and velocity meters may be used in heated and chilled water circulation systems.

11.5.2 Sub-Metering

The sub metering of buildings provides the operations and maintenance transparency necessary to enable more efficient management of energy and water resources. In addition, sub metering can drive behavioural change related to energy conservation and advance real-time building interaction with the Smart Grid. Each of these potential benefits can dramatically improve building performance and lead to reduced resource consumption.

While sub meters by themselves have no direct impact on resource use, the data they capture informs real-time energy and water performance, can pinpoint performance variations over time or relative to other buildings, feeds into an EEMS that drive continuous operational improvements, and provides the information needed to encourage behavioural and operational changes by building operators and occupants.

Sub metering systems shall be carefully designed to meet stated operational criteria and objectives, i.e., data analysis and operations management requirements must guide sub metering hardware and software selections and specifications of the system configuration. sub metering designs shall include the specification of the level of sub metering required, the types of data to be collected and used to reduce consumption, and the software, analysis, and communication tools necessary to manage energy usage by systems, to provide information to operators or occupants and/or to control the building's energy and water use directly. Sub meters shall interface with the building automation system, not only making it possible to integrate the control logic with the meter, but also providing access to the building automation system tools for trend analysis, report generation, and user information display

All building core systems (air, water, cooling, etc.) shall be isolated, metered, tracked and trended through the application of current sensors and system technology. The metering system shall be capable of monitoring and recording utility used for chilled water, air flow, domestic water use, and electricity use on a floor-by-floor and tenant-by-tenant basis.

Dedicated temperature and flow sensors shall be installed on the primary chilled water main headers to monitor and measure consumption in a traditional unit of heat (example -MJ/GJ) and demand.

Electrically, meter boxes should have spaces for future meters, and core building load should be metered on a separate meter.

System selection for leased tenant locations shall be designed to operate independently of the base building system so that only the utility connections need to be sub- metered.

Thermal Energy meters that display the consumed energy quantity in MJ / GJ shall be provided where chilled water is supplied to leased tenant space from a central location.

Measurement, verification, recording of and feedback from the building HVAC system should be provided. Verification of air quantities, ventilation rates, and air quality shall be required as specified in the OPR.

11.5.3 Meter/Sub-Meter Performance Metrics and Attributes

Performance measures for meters and sub-meters shall include but not limited to the metrics listed below:

- Accuracy
- Precision/repeatability
- Turndown ratio
- Ease of installation
- Ongoing operations and maintenance
- Installation versus capital cost

11.5.4 Communication Networks and Data Storage Requirements

Regardless of the meter type, once sub-metering data is collected it must be transmitted via a communication network to be processed, stored, and used. The type of sub-metering plan with its requisite data requirements will define the communication network appropriate for the application. The options for automated metering communications include phone modem, local area network (LAN), building automation system, radio frequency (RF), and wireless network.

Metered energy data shall be analysed at regular intervals (e.g., every 15 minutes) and may incorporate other relevant information:

- Account or location
- Usage points (e.g., lighting, HVAC)
- Kilowatt-hours of consumption
- Highest kilowatt demand
- Voltage
- Real and reactive power
- Hourly weather information (temperature, humidity)
- Local currency values of consumption, peak charge, access charge, taxes, and delivery charges
- Calculated values for load factor, per square foot or per square meter use
- Heating or cooling degree days

For water distribution systems, metered data shall include:

- Water flow rate
- Peak flow rate
- Pressure
- Temperature
- Hourly weather information (temperature, humidity)
- Location

The above-mentioned information shall be integrated with software analysis capabilities. A building’s advanced meter, sub-meters, and building automation systems shall all provide data to an EEMS which shall have analysis tools ranging from spreadsheets on local computers to integrated systems with analysis capabilities, trend identification, report generation, and graphical displays.

Dashboards with both energy and water use applications, equipped with features and interfaces that are tailored to the needs of the facility operator shall be provided. They shall include real-time load profiles and baseline comparison profiles that correct for weather and other external conditions, so that the facility operator can accurately determine the impact of energy conservation initiatives.

Figure 11.5 Utility Metering Layout

SECTION 12.0

Plumbing and Sanitary Systems

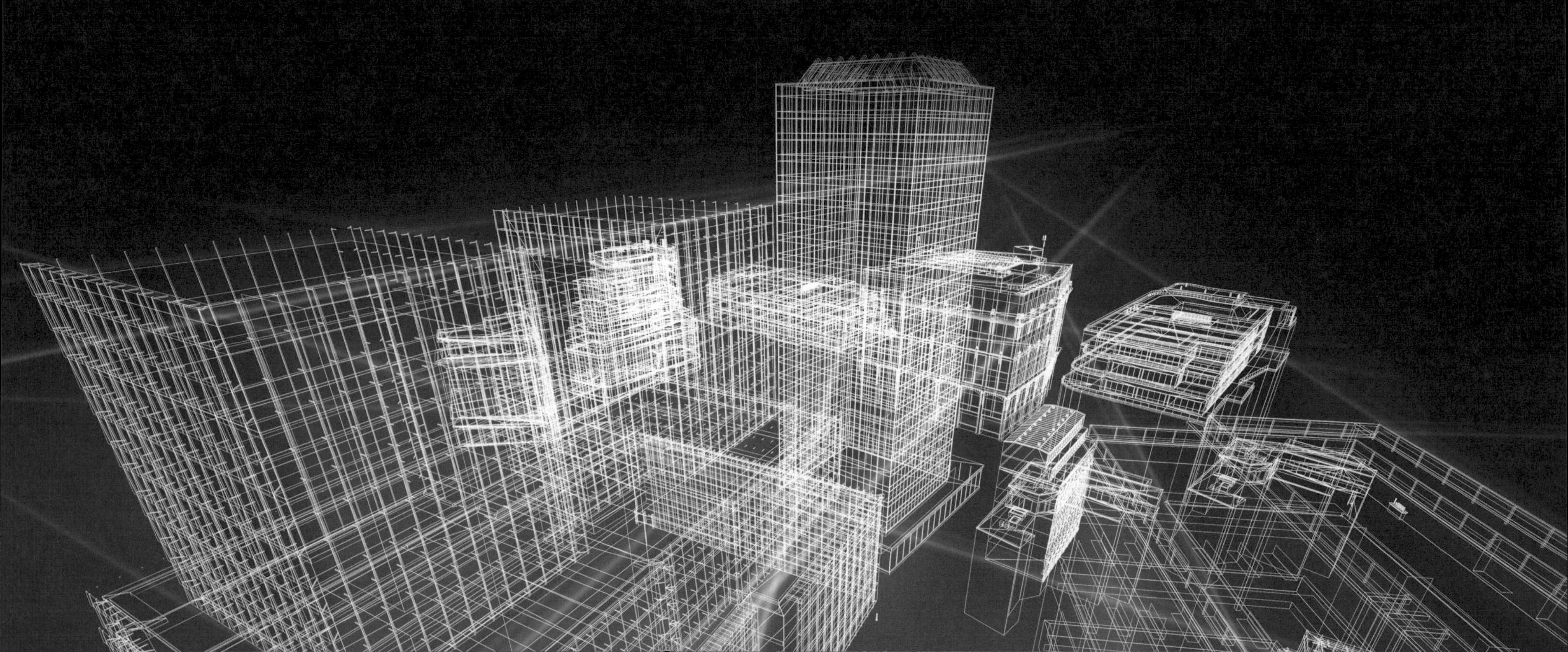

12.1.1 General

The design team shall determine the overall design solution that addresses the technical, physical, and economic aspects, that complies with recognized International Standards, with the aim of providing safe and reliable hot and cold water, and sanitary drainage systems for the comfort and health of Facility occupants. In particular, all current recommendations, including Health and Safety at Work Act 1974 for prevention of Legionella should be complied with as a minimum.

12.1.2 Domestic Hot and Cold Water

Many variables should be considered during the engineering of domestic water systems, and several design solutions are to be considered by the design team. The water pressures vary at different locations throughout the building, and therefore, must be considered in system layouts and when selecting equipment and pipe materials. Energy efficiency, space allocations, economics, and acoustics should all play important roles in a successful project delivery.

The hot and cold-water distribution systems shall be sized using 'demand units' based on CIBSE, ASHRAE or the latest version of International Building Code (IBC). In addition, the facility's occupancy, and group usage for combined meetings, shall be considered.

An adequate, safe, healthy and dependent water supply should be provided to the property by the local utility supplier. Where this is not available, a suitable pumping/ treatment system should be provided as part of the design to ensure the available supply meets with World Health Organization (WHO) requirements, as a minimum. Additional treatment should be applied to ensure the supply is neither corrosive nor scale forming. All water supplies to heating plant should be softened if required to meet International recommendations. Cold water supplies should only be softened on Owner approval.

Water pressures should be established for all points in the domestic cold and hot water systems. The residual flow pressure required at the fixture could vary depending on the fixture function and design. There should be sufficient pressure at the fixtures to assure the user of a prompt and adequate flow of water. Pressure zoning of domestic cold water should be closely coupled to the design of the hot water system.

Storage should be provided as indicated in this section, or as determined by the availability from the utility supplier, with a minimum of 24-hours for maintenance purposes. If dual 100% reliable supplies can be demonstrated, then storage may be reduced – this is subject to Owner approval. Cooling should be provided to potable water in hot climates to achieve the specified supply temperature.

Several configurations of pipe distribution systems should be employed to distribute hot and cold potable water to the facility. Separate potable and non-potable supplies should be provided to ensure water quality in accordance with WHO requirements. All residential apartments, public washrooms, food preparation areas and staff areas should be provided with potable supplies.

Where adequate flows/pressures cannot be achieved by gravity supplies, constant pressure booster sets should be provided, in accordance with Section 10.4.2 ASHRAE Standard 90.1.

System pressure should be maintained by the use of variable frequency drives (VFDs). Remote mounted pressure sensors or advanced control logic shall be used to allow pump speed reductions at low flow when there is less friction in the piping.

The design team shall investigate several design solutions for domestic hot water supply (DHW) systems. Basic design choices such as conventional heater tanks, solar water heaters, energy recovery heaters etc should be investigated.

In commercial buildings that use limited hot water at the core washrooms, hot water may be served locally, either with instantaneous electric heaters at each group of washrooms, or with small electric heaters that may serve a grouping of floors. Where large, central hot water systems are required, such as for the residential apartments, pressure zoning should be fully addressed.

DHW should be circulated at 43°C, and thermostatic mixing valves or automatic temperature compensating valves must be provided to prevent scalding. Hot water should be discharge at the outlets at design temperature within 10 seconds of opening the outlets. Materials that encourage the proliferation of Legionella (e.g. "rubber" flexible connections) shall not be used. Provision should be made to achieve a water temperature of up to 55⅓C at all outlets, including those served by thermostatic mixing valves, to allow for hot flushing of the outlets, either through the mixing valves or via a by-pass.

Heat sources shall be gas, electricity or fuel oil fired boilers. If the facility is served by steam, this may be used as the heat source for DHW systems. Because of the relatively low final temperature needed, DHW could be provided by solar energy equipment, heat pumps, and heat recovery devices.

For energy efficiency, a combination of central and distributed DHW may be investigated. Energy saving microprocessor control devices that vary the supply temperatures of hot water so that the hottest water is supplied at the busiest hours should be considered. Separate duplicate hot water storage tanks should be incorporated for commercial kitchens/laundries at a higher temperature.

12.1.3 Sanitary Drainage

The foul sewer system should be designed in accordance with local bylaws and the latest version of the IBC, to be self- cleaning, self-venting and fully suitable for the purpose. Multistory office buildings are generally designed with a central "core" that contain shafts for plumbing.

Gravity flow in horizontal branches or vertical stacks shall be sized using Drainage Fixture Units (DFU), as described in the latest version of the International Plumbing Code (IPC) or equivalent standards.

Grease traps should be provided for all commercial kitchen areas, and installed external to the building, unless proprietary chemical types are used. They should be in a location accessible by vehicles for cleaning. Interceptors should also be provided to prevent the discharge of oil, grease, sand and other substances harmful or hazardous to the public sewer, the private sewage system or the sewage treatment plant.

Sizing of the grease interceptors shall follow the IPC requirements or equivalent. Grease interceptors shall be equipped with devices to control the rate of water flow, so that the water flow will not exceed the rated flow.

Manholes should be external to the building and at every junction and change of direction.

12.1.4 Storm Water Drainage

Based on the natural characteristics and irregularities of the site, the storm water runoff design should include an investigation of the vegetation and compactness of the soil. The drainage system shall be designed to slow runoff velocities and volumes to encourage infiltration. Local rainfall charts shall be used to select a suitable design.

Storm water management choices should include pervious pavers, retention ponds, underground chambers, and rain water harvesting. Rain water harvesting should have provisions to collect water from off roofs and reuse that water. Storing the rainwater on site eliminates a major portion of the storm water drainage infrastructure. Such harvested rainwater should be used for non-potable uses such as landscape irrigation but can also be a source of water for flushing toilets and laundry, through integration to the internal plumbing of the facility.

12.1.5 Water Treatment

Water treatment should be installed if any of the following problems exist:

- Hard Water – softener only to be applied to hot water services, unless levels such that cold services would be affected.
- Alkalinity
- High bacteriological count
- Discoloration
- Excessive organic growths
- Corrosion due to soft water - The water supply should be neither scale forming nor corrosive.

Particular consideration should be given to the following systems:

- Chilled water
- Condenser water
- Domestic hot and cold water
- Kitchen wash up and laundry equipment
- Spray coil humidifiers (if used)
- Steam Boiler (Laundry)

Treatment for Legionnaires Disease in all systems should be in accordance with HSE (UK) Legal series L8 document – "Legionnaires Disease – The Control of Legionella Bacteria in Water Systems.

12.1.6 Gas Supplies

Provide an incoming gas supply to enter the property in a location that allows unobstructed access by the utility company, in a back of house location. Gas services should be installed along the shortest practical, available and accessible route that is not subject to undue stresses, hazardous conditions or obstructions.

Bulk storage of LPG, Butane or Propane shall be used, based on 6 weeks' supply at maximum use. Refer to NFPA 58. Provide separate metering for each area served. Provide automatic gas detection with solenoid shut off valves connected to the Fire Detection and Alarm System for each area served.

SECTION 13.0

Vertical Transportation

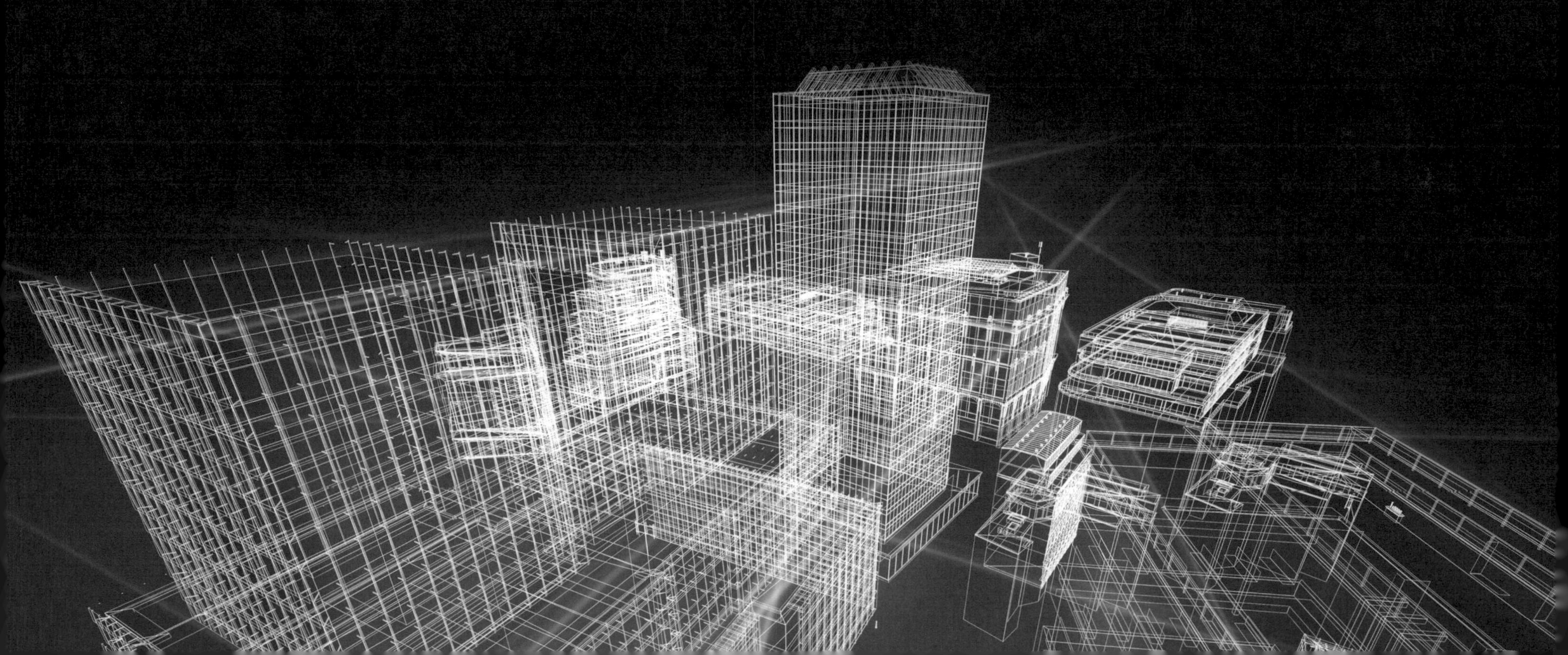

13.1.1 General

The design team should ensure that the vertical transportation devices (lifts and escalators) which provide mechanical movement of people, will receive and dispatch passengers efficiently and effectively. Various circulation elements and their characteristics should be investigated, prior to consider their locations. Lifts and escalators should be considered together – as a single solution in vertical transportation – for the particular facility being designed.

The location of lifts shall be such that they are easily accessible and convenient to circulation routes. When additional lift banks are provided, every effort shall be made to locate them along the same major circulation paths that serve the existing lifts. Adequate area for circulation and waiting of staff, visitors, and equipment shall be provided. Lifts shall be located to provide positive separation between Residential, commercial operations, and service traffic flows.

When exploring the various elements, minimize the movement of people and goods; prevent clashes between people and goods and prevent bottlenecks. Therefore, the location and arrangement the passive circulation (corridors, portals etc.) and the active circulation elements (passenger lifts, escalators) should take account of:

- The location of entrances and stairs
- The location of lifts and escalators
- The distribution of occupants in the building

Generally, escalators should not lead directly off corridors, but should accessed from landing and lobby areas, where people may wait without obstructing a circulation route. It is particularly important that the boarding and alighting areas adjacent to an escalator are not part of another circulation route.

Lift lobbies should not be part of a through circulation route, either to other lifts, or other areas in the facility. Lobbies should be provided that are dedicated to passengers waiting for the lifts. The ideal lift lobby size should be one which could accommodate one full car load of passengers waiting and allow the simultaneous disembarkation of one full car load of arriving passengers.

13.1.2 Elevators/Lifts.

Passenger Elevators/Lifts Specifications for vertical transportation installations shall provide that at the time of the completion of the Facility, and the beginning of its operation, the Elevators/Lifts and/or escalators shall be completely adjusted and checked out, approved by local authorities and in god operating condition. All Elevators/Lifts equipment shall be designed, furnished and installed to comply with all local prevailing Elevators/Lifts, electrical and building codes, including revisions and/or supplements to the date of the contract. In the absence of an applicable code, the Elevators/Lifts equipment shall be in accordance with the most current revision and supplement of ANSI A17.1. Compliance with local prevailing handicap codes is mandatory. Americans with Disabilities Act Guidelines (ADAG), shall establish design guidelines in the absence of an applicable code.

Design Criteria

In residential buildings passenger lifts are to be located in a central core to provide efficient and flexible utilization. The objectives are to provide waiting passengers with quick and convenient access to the lifts.

The apartment building vertical transportation requirements should be based upon the population, location and type of tenant. A building with many families would experience a school-hour peak: buildings in the city with predominantly adult tenancy will exhibit evening peaks due to the homecoming working group and outgoing dinner traffic.

Self-service collective control would apply generally, with provision for attendant control in luxury apartment buildings.

The objective of vertical transportation in a retail shopping complex would be to transport persons to specific levels, and partially to expose the customers to displayed merchandise. The design team should investigate the possibility of relying heavily on escalators, with adequate number of lifts provided for use by staff and handicapped persons.

A traffic analysis shall be made based on the following guidelines.

Two basic types of specifications for elevator equipment may be applied. They are the prescriptive (equipment based) and performance (outcome based) approaches. Due to the advent of pre-engineered, pre-manufactured systems, performance specifications may be considered.

- Sytem handling capacity - 5-minute peak period demand (Percent based on 100% occupancy)
- City Center Office Building 10-12%
- Luxury Apartments 5-7%
- Residential 6-8%
- Average interval - Two-way morning and afternoon peak periods:
- City Center Office Building 15-24 seconds
- Luxury Apartments 50-70 seconds
- Residential 60-80 seconds
- Performance

System Response Times: Percentage of lobby calls answered under one minute

- 50% in 30 seconds or less
- 95% in 45 seconds or less
- 100% in 60 seconds or less

Conference/Meeting Facilities Lifts:

- Special function occupancies - square metres per person:

    o Meeting Rooms 1.5

    o Ballroom 2.0

    o Exhibition 2.5System handling capacity:

If the ballroom is served by a grand stairway, 50% exit capacity should be assigned to that stairway.

Service Lifts:

In office buildings, approximately one service lift is required per 10 passenger cars or alternatively, one service lift for every 27,870 m^2 of net area. Service lifts should be 2268kg or larger without dropped ceiling, and if used for passenger service, should be equipped with wall pads. An oversized door (1.2m – 1.4m) should be provided to facilitate handling of furniture items. The higher the commercial profile of the facility, the more service lifts should be incorporated. A larger size service lift may be required to transport function displays to ballroom or exhibition rooms.

The service lifts should be large enough to handle bulky furniture, and should therefore be at least 1814kg with a 1.2m door and high ceiling. The lift hoist ways should be isolated from sleeping rooms by lobbies or other space.

Freight Lifts:

Freight lifts shall be specifically designed for transportation of freight only. freight elevators shall be provided, where appropriate (ballroom, banquet kitchen, exhibition areas etc.). Factors to be considered in freight lift selection, in addition to the load to be transported per hour, are size of load, method of loading, travel, type of load type of doors, and speed and capacity of cars.

Dependent on the lift speed, a geared type traction machine or a hydraulic unit may be used. The preferred system of control is collective, with a variablevoltage variable frequency (VVVF) For a low-rise installation, a hydraulic unit could be selected.

Freight lift cars should be built with heavy-gauge steel with a multilayer wooden floor, the entire unit designed for hard service.

13.1.3 Escalators:

Escalators could be configured in a crisscross arrangement or a parallel arrangement. The designer should provide adequate lobby space at the base for queuing where anticipated. At the exit terminus, the escalator should discharge into an open area with no turns or choice of direction.

In meeting and conference facilities, when a meeting/ conference or special functions are located above or below the Entrance Lobby, escalators may be considered to accommodate peak traffic conditions and provide positive separation from the passenger lifts for security purposes.

To meet ADA requirements, elongated newels with a minimum of two horizontal treads before the landing plate should be provided. All escalators shall comply with sections 6.1.3.5.6and 6.1.3.6.5 of ASME A17.1, and shall have a clear width of 813mm minimum.

Escalator Handling Capacity: A full auditorium is to be transported to the Main Egress Level in a maximum of 30 minutes. Assume a maximum of 30% of the population will use convenient stairs or lifts. Maximum linear speed of the escalator shall be in accordance with the safety code (ANSI/ASME 17.1). Preferred Escalator Application (subject to traffic analysis for each specific project.

Population of Typical Mixed - Use Buildings

Building Type	Net Area
Office Building	m^2 per person
Multiple Tenancy	
Normal	10 - 12
Prestige	14 - 23
Parking Garages	19
Single Tenancy	
Normal	8 - 10
Prestige	12 - 19
Residential	Persons per bedroom
High-rental Housing	1.5
Moderate-rental Housing	2.0
Commercial Spaces	
Enclosed Shopping Mall	Consult the local code
Large Stores	2.8 – 4.65 m^2 per person
Restaurant	1.4m^2 – 1.7m^2 per seat
Food Court	1.5m^2 – 1.7m^2 per seat
Counter Service	1.7m^2 – 1.85m^2 per seat
Auditorium (tiered)	Traditional seating 0.75m^2 per person
Auditorium (tiered)	Continental seating 0.88m^2 per person

SECTION 14.0

Audo/Visual (A/V) Systems

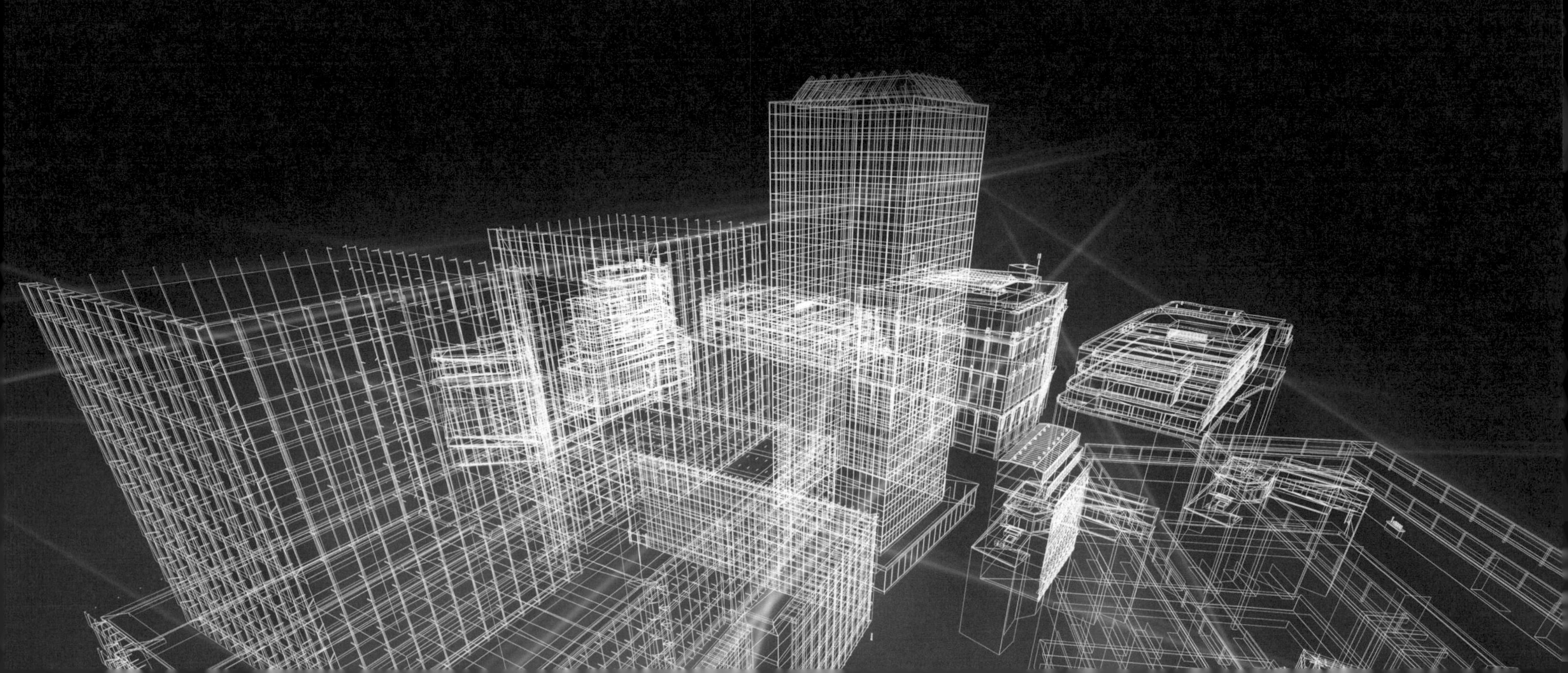

14.1.1 General

This document is a guideline to the professional engaged by the Owner for the design of Audio-Visual systems. It is to be used in the preparation of contract documents for the construction Mixed-Use Building A/V facilities. The intent is to establish minimum and sufficient technical criteria, and typical functional requirements or "standards" for each facility.

These guidelines are not to be considered absolute; special considerations will arise in each facility. In addition, the fundamental technological dependency of the Systems design afforded by technical advances may suggest alternative approaches. While the very existence of these guidelines suggests a desire for uniformity, alternative approaches may be considered.

Due to the variation in design of the facility's this guideline applies to, reflecting the diversity of architectural design concepts, the needs of the tenants, deviation from these standards may be required. Where such deviation is required by the acoustical or visual environment, the A/V consultant shall advise the Owner, the nature and impact of the deviation. Deviations should be limited to providing equivalent or better perceived quality to the tenant and should only be pursued in order to maintain or improve the quality of the systems by cooperation with the architectural features of the facility.

The overall intent is to provide a design of maximum flexibility and quality, while reducing the amount of user interface required for normal operation, to a minimum.

14.1.2 Coordination:

I. General Coordination; At some point prior to the preparation of Contract Documents for construction, documentation would be prepared for review by the Al Fardan Commercial Properties, and as appropriate by other design team members.This documentation would take the form of both concise written descriptions of each system, its implementation, and as well schematic architectural drawings depicting the locations of the various devices and equipment. This documentation shall serve as the first step in an ongoing dialogue, culminating in the finished construction.

II. Coordination with the Design Team;

The "design team" will typically comprise the Architect, Interior Designer, Kitchen Consultant, Electrical Engineer, Mechanical Engineer, Structural Engineer and Civil Engineer.The team usually will include a Laundry Consultant, Landscape Designer, IT Consultant, Audio-Visual & Security Consultant, etc. The nature and scope of the "Audio-Visual Systems" work is such that it will impinge on the work of most of the other design team members, if not all. It is vitally important to identify the impact of this work on the work of others and to establish appropriate communications with the respective team member at the earliest. Each team member shall be made aware of the specific electrical device or load, heat load, architectural requirement, space requirement, structural requirement, visual impact, finish, etc. as appropriate, relative to the design of specific audio- visual systems and equipment. Ultimately, the successful coordination of these disciplines will realize appropriate annotations,where necessary,in the documentation provided by each team member acknowledging the associated work by others, and directing the various construction trades to the appropriate documentation. Such efforts span the most fundamental review of architectural development relative to ceiling heights and sightlines, projector sightlines, adequacy of clear heights for given presentation activity, space planning relative to the implementation of control booths or interpreter booths, etc.

III. Role of the Audio-Visual Systems Consultant;

The role of the Consultant has been generally stated in the preceding three paragraphs. More specifically, the Consultant is responsible for several areas of expertise; preparation of Design/Development drawings, preparation of Contract Documentation, including Plan Drawings, Reflected Ceiling Plans, supportive drawings, and Specifications, Coordination with the Design Team including the Owner, Review and Comment of Contractor Submittals, and Site Inspections, which would include preliminary and final review of the various Systems. In as much as the various Systems for which the Consultant is responsible are usually purchased by the Owner, it is essential that the Consultant's Plans and Specifications not require any supportive documentation issued by the Architect. This operative technique should be coordinated and clarified with the Architect.

IV. Related Work by Others:

- Define extent of conduit provisions explicitly within the Audio/Visual Bid Documents and Audio-Visual Contract Documents.
- Coordinate specifications requirements with the Electrical Engineer such that the Electrical Contractor provides the conduit, conventional electrical style back boxes, pull boxes and pull strings indicated in the Audio/Visual Contract Documents and installs loudspeaker enclosures, specialty boxes specifically indicated as furnished by the A/V contractor, and shown on the Documents.
- Coordinate the heat load with the HVAC Engineer at any equipment locations where the associated heat load is in excess of 295 watts, or where direct connection from HVAC system to audio visual equipment is required.

V. Design Provisions:

- Define grounding and shielding requirements within the audio-visual bid and contract documents.
- Define portable equipment (patch cables, microphones, assisted listening systems, etc.) that will be essential to basic operation, to be furnished by the A/V Contractor. Include sufficient equipment for the Facility Management staff to hold department meetings and other internal events. All A/V equipment shall fully comply with latest version of UL 813 standards.
- Review impact of projection screen sizes, reasonable audience sizes, observer sight lines, projection sight lines, audio visual equipment connection method and location, portable stage equipment and theatrical lighting provisions at recommended head table or "stage" locations for meeting spaces with Owner and Architect, and recommend any remedial actions to coordinate these factors with the design goals of the facility.
- Review finish materials and surfaces with respect to acoustical considerations directly impacting audio systemperformance. This should be considered as a guideline for auditorium, ballroom and meeting rooms, and as a mandatory design consideration for areas provided with a fixed seating arrangement.
- In areas where permanent or regular Videoconferencing is part of the program, the placement and acceptability of the surface materials

and finishes shall be dictated by the acoustical requirements of the Videoconferencing systems. The A/V consultant and interior designer shall coordinate the efforts to minimize the reverberation in the room as well as eliminate all detrimental reflections.

VI. ADA/DDA Compliance

All areas with fixed seating arrangements shall be provided with a Hearing Loop assistive listening system, including receivers for use by the hearing-impaired. The assistive listening system (ALS) shall cover the entire seating area. The ALS system shall conform to IEC 60118-4:2006 standards. Only Loop Drivers that meets measurement standards set forth in IEC 62489-1 shall be provided. All areas provided with sound reinforcement systems shall have assisted listening systems available upon reasonable notice.

In areas provided with permanent assisted listening systems, review the cost with the owner, to provide sufficient bandwidth/radiating power to accommodate simultaneous interpretation of up to four languages with the original signal.

Include any portable equipment required for ADA/ DDA compliance in the list of portable equipment to be furnished by the A/V contractor.

14.1.3 Infrastructure Cable Plant Design

TCP/IP Communications: The designer shall provide a future-ready, scalable, high-performance structured A/V solution, and Implement code-compliant, standards-based solution, for a custom designed architectural cabling infrastructure installation for voice, data, and wireless. All necessary racks, patch panels, and cable management systems shall be provided.

Cabling circuits for the A/V systems shall conform to and comply with the latest version of National Electrical Code (NEC) Article 640 and 725. Where Data (Ethernet) circuits are required, these cabling systems shall conform and comply with NEC Article 725.The designer shall take all necessary precautions to prevent signal crosstalk.

In an A/V system there may be only one Ethernet connection with the remaining cables supporting audio, component video, broadband video, HDMI, RS232 serial, USB, and DVI signals.

The A/V system designer should consider the separation of cables or signals from a performance perspective. Signals of different voltage levels should be separated to avoid signal cross talk. Microphone level (mV), line level (2 V and less), and speaker level (25 to 100 V) signals should all be separated.

The structured solution must include HDMI ports. HDMI connectivity should be on the wall-plate leading from a conference table to the display, and the connection needs to support the resolutions anticipated supporting over the life of the installation. If the Owner require certified Hi-Speed HDMI(e) performance that supports up to Ultra high-definition video with 4K and even 8K video resolution, the system should be adaptable to emerging technologies for several years.

The A/V infrastructure should feature digital DisplayPort connectivity. Computers, tablets, convertible devices and USB Type- C enabled products will all come with advanced DisplayPort connectivity, and VGA is fully adaptable to a DisplayPort environment. A DisplayPort connection should be on the wall-plate from the conference table to the display.

Since USB, HDMI and DisplayPort-has tight length restrictions, HDBase-T or optical media conversion may be adopted as a standard for whole-home and commercial distribution of uncompressed high-definition multimedia content.

Where required, a stationary RF wireless audio distribution system should be provided.

Cabling Accessories: It is essential to have the right A/V connection in the right location to provide the flexibility to conveniently connect and disconnect various devices.

Connectivity and Delivery Systems solutions shall include, but not limited to: Fire rated floor penetration accessories that accommodate power, data and A/V within the same space. Fire rated floor boxes that deliver flexibility and maximum capacity for all A/V installations. The boxes should support multiple HDMI and A/V connections in numerous configurations.

Since interactive solutions will be an element of the space, it should be included it in the design. Integrate A/V into a conference table top, which allow convenient access to network drops, AC power sources and A/V displays. These units shall have open architecture design that allows for both analog (VGA), digital (HDMI) connectivity to be configured, Wall Boxes that have A/V, data, power and building management system (BMS) residing side-by-side in the same box.

Provide A/V ceiling boxes that mount into the ceiling grid, and provide adequate space for active and non- active A/V gear, power connections and cable storage and management.

14.1.4 Audio/Video Systems

I. Background Music (BGM) Systems:

- ***Function;*** To reproduce musical program material for non- critical listening and environmental enhancement.
- ***Application;*** Provide in all public circulation areas; restaurants, washroom facilities, the health spa and swimming pool, and public lifts. Background music programming would additionally be provided, typically as part of the local sound reinforcement system, within exhibition, meeting, banquet and ballroom areas.
- ***Description;*** The System would comprise a high-quality music reproduction system that could receive signals from portable devices, music servers, disc players, streaming music services etc., as a continuous program material source; multifunction preamplifier with integral digital-to-analog converters (DACs) configured and controlled via software. The system shall be a dedicated plug-and-play audio control system, which could control individual spaces from the front panel, making it possible to set-up and monitor all spaces from a single location.

Emergency and Life Safety Operation: The sound systems described here are not intended to be used as life- safety systems. Alarms or emergency paging shall occur through other systems specifically designed for such use. The sound quality and dependability of the equipment specified herein usually substantially exceeds that of the equipment available that is approved for life safety use. As such, the fire authority may request access to the systems as a supplement to the approved systems. Such access will be accommodated on request, however, under no circumstances will equipment that is not approved for life safety use be the sole means of communication. If the life safety system is interconnected to the A/V system, provide all volume controls and sound reinforcement systems with a non-interruptible path for the audio signal to the listener.

It is required that all non-essential sound systems become inoperative in the event of a life-safety alarm. The A/V consultant shall coordinate with the electrical engineer to de-energize the power circuits to the equipment rack to comply with the ordinance. In this circumstance, provide uninterrupted power supplies for all sensitive equipment that would be damaged by sudden shutdown (it is essential that all output from power amplifiers ceases immediately in this circumstance) and provide sequential power control to energize the appropriate equipment in a safe manner when power is restored. To avoid unnecessary cycling of the equipment as well as preventing shutdown of power at the time of programming of any computer controlled equipment, if acceptable to the local fire officials, the power shall not be disabled during testing of other alarm circuits until final inspection.

(a) **Basic Sound System Performance;**

- This System would be capable of producing a continuous sound pressure level (SPL) not less than 15db greater than the average ambient noise level or 90db SPL, whichever would be greater, as measured 170 cm above the finished floor. Frequency response would be +3db from 80 to 10,000 Hertz, relative to 1,000 Hertz. Uniformity of coverage would be +3db relative to the level measured on the axis of any one loudspeaker assembly, measured 170 cm above the finished floor within the area of System coverage, at 1,000 Hertz.
- Coordinate placement of loudspeakers in corridors and primary travel paths through public areas to provide a maximum level deviation of +2dB, - 4dB from 100 Hz to 4000 Hz in the direction of travel. This parameter is to be measured relative to the overall frequency response criteria above. The desired result is that the listeners do not perceive the level deviation as they travel through the area, however, the travel path can be in a "fringe coverage" area where high frequency coverage is compromised equally throughout the travel path

(b) **Appearance and location of panels shall be as follows:**

- Input/output panels shall be mounted at the same height as AC power receptacles.
- Volume control panels shall be mounted at the same height as lighting controls unless concealment provisions dictate otherwise. The control labeling must be readable and the controls readily accessible.
- All panels visible to the public should complement the interior design.
- Coordinate receptacles and control panel locations with the wall construction, and other receptacle assemblies and panels as indicated by the Interior Designer on the elevation drawings. Coordinate and verify the subsequent documentation of the nature and location of all electrical power provisions for audio visual systems on the electrical plans. Include the associated grounding, electrical load and demand factor in coordination efforts.

(c) Appearance of Loudspeakers

- Distributed speakers shall usually be mounted in the accessible areas of the ceiling wherever possible. Spacing should be arranged to provide even coverage throughout the audience area while maintaining an aesthetic or unobtrusive relationship with the ceiling architecture and other elements.
- Loudspeaker grilles should be white or painted to match ceiling.
- Coordinate the location of loudspeaker devices with the acoustical environment, layout at partitioned spaces shall accommodate the various room configuration without disruption of sound coverage, when the partitions are removed and replaced.
- Loudspeakers exposed to direct contact with rain, sprinklers, or hosing overspray shall have diaphragms and baffles impervious to water.
- Utilize stainless steel or plastic loudspeaker grille products with minimum open area of not less than 40%.
- Loudspeakers exposed to high humidity or foggy environments shall be provided with moisture resistant diaphragms and loudspeaker baffles. Utilize stainless steel or plastic loudspeaker grille products with minimum open area of not less than 40%.

(d) Location of Sound System

The sound system equipment rack(s) for the central background music system shall be in the Data Centre, for operation by facility staff.

The equipment rack shall be positioned so that it may be serviced front and rear without being moved.

Equipment and controls which must be available for operation include:

- Volume, source and combining controls for all zones specified as being controlled from the central rack.
- Background music source equipment.
- Music source selection facility.
- Patching provisions.

II. Distributed Loudspeaker Sound Reinforcement Systems

- Function; To reinforce live presentations including spoken word, singing, musical instruments, etc. Implicit in the word "reinforce" is that reproduction is an extension of the live program, and hence, sonic accuracy or fidelity is a fundamental concern.
- Application: Provide in all meeting rooms, function rooms, board rooms, and ballrooms rooms over $85m^2$ area in size. Systems may additionally be required in lobby lounges either to augment a "Central Loudspeaker" sound reinforcement system, later described, or as the sole means of sound reinforcement where room geometry precludes the latter type of system. Where contiguous rooms, each over $85m^2$, are separated from one another by folding partitions, each room shall be provided with a separate system, as described below, and later under "Audio Control Systems". Where contiguous rooms are separated by folding partitions and accumulative area is more than $85m^2$, a single system would be provided to support use of the combined space.
- Description; The equipment shall be rack or shelf mounted chassis. This chassis shall accommodate the main audio input card and output cards.

- The Background Music System shall work in conjunction with all distributed music systems (AEI, DMX, Muzak), as well as any audio source (CD, DVD, etc.), and allow a selection between 10 stereo audio sources from a wall plate or remote wireless device. The digital stereo audio and control signals shall be transmitted and distributed utilising a single CAT-5 cable The Conference & Meeting Manager's software option shall be Windows based, and able to mix the audio systems, control the in-room panels and adjust volume in banquet function rooms. The wall-plates in banquet function rooms shall have the capability to be home run and/or daisy chained.
- Location of Sound System; The equipment racks for the ballrooms and small meeting rooms utilizing critical controls installed in rack or shelf mounted chassis, shall be in close proximity to these rooms.

III. Central Loudspeaker Sound Reinforcement Systems:

- Function; To reinforce live presentations including spoken Word. Implicit in the word "reinforce" is that reproduction is an extension of the live program, and hence, sonic accuracy or fidelity is a fundamental concern.
- Application; Provide in all meeting rooms, boardrooms, ballroom and other presentation facilities, architecturally configured with a single defined podium, platform and/or stage, where the room geometry would be over 110m² in area.
- Description: The System would comprise a complement of high-level, low and high frequency loudspeaker assemblies and at a minimum, a local level control, local or centrally located mixer/digital processor, and power amplifiers.

IV. Conference Microphone System

- Function; To provide an efficient 100% wireless infrared (IR) conference system featuring infrared communications and microphone units. Additionally, the system shall have sophisticated functions, such as simultaneous interpretation for a maximum of one native language and five foreign languages, automatic turning of camera following the switching of speakers. The system shall be equipped with full complement of Conference Manager Software.
- Application; Provide in all boardrooms, and other presentation facilities, architecturally configured with a single defined conference table or other appropriate seating facility.
- Description: The System would comprise a master control unit, microprocessor controlled interpreter desk units, delegate units, wireless hand-held microphones, IR transmitter/receiver etc.

V. Monitoring & Control System:

Ballrooms and Meeting Rooms shall be provided audio/ video control units in a shallow recess. The control units shall consist of remote scene select wall stations operable by touch screen controls.

The lighting dimming systems shall communicate with the audio/visual system to provide an integrated scene in each area.

Remote programming jacks shall be available for each monitoring and control system to set and adjust scenes, as required.

The recess where the control units are located shall be well ventilated, and supplied with power and network as well as signal and control cabling.

14.1.5 Function Rooms and Boardroom

All meeting rooms shall be equipped with a Digital Media Presentation System.

(a) **System Architecture**

The Digital Media Presentation system shall be a single central switching and control unit. It shall use a touch screen to carry out functions such as powering a data projector, selecting sources or controlling the audio levels. It shall provide the ability to monitor and remotely control equipment in the system as well as providing maintenance and security alerts.

Lighting system functions shall be accessible by the A/V Automation Control Processor.

The Central Switching and Control Unit shall include an integrated microprocessor based control processor.

(b) **Conference Equipment**

An Interactive Board for boardrooms, an LCD or Plasma Screen, or a ceiling recessed tensioned electric screen and mounted data projector for all other meeting rooms, shall be provided. The AV cabinet and a control panel shall be installed in a well ventilate recess, and supplied with power and network as well as signal and control cabling.

Visibility being critical consideration shall be made for different aspect ratios, some of it in 4 x 3, some of it 16 x 9, to accommodate the needs of users who come with a variety of source content, which itself, can be anything from a website, a PowerPoint presentation or electronic writing. The room should be accessible, and designed with wheelchair users in mind as well as being comfortable for keyboard use whilst standing.

(c) **Touch Screen Control**

Provide a touch screen that enables the end-user to control all portions of the A/V system, from the audio volume to the projector on/off and inputs to the projector screen up/ down.

Each room shall have a stand-alone unit and one built into the wall.

An iPad application shall enable one to operate the complete Control System remotely over a wireless network.

(d) **Projector**

The projectors shall have two VGA inputs for a PC or laptop, as well as a component and/or composite input for a DVD or camera feed.

The projector shall be mounted on a Smart-Lift Automated Projector Mount, and recessed in the ceiling space. The operating system shall be control options of a low voltage controller, and internal and external wiring options, and accepts 5-30V AC/DC for triggering commands.

(e) **Switcher/Scaler**

Provide a video switcher/scaler, which will take a variety of video inputs from a computer, DVD player, camera, etc. and convert them all to the same size and display format.

(f) **Network**

The network requirement associated with a meeting space shall comply with the requirements described in section 12.3. This assumes wireless network provision in the meeting space for laptops, and allows some room for expansion of the system if required. The wiring requirement associated with signal and control in a lecture environment shall be concentrated in the portable lectern with additional cables being run to external devices such as the data projectors, cameras, additional microphones, loudspeakers etc.

14.1.6 Ballroom

The ballroom shall be equipped with quality distributed sound reinforcement system suitable for speech reinforcement and background music. All speakers shall be recessed ceiling type, inset wall mounted or of similar configuration.

Amplification systems shall be rated such that the number of loudspeakers connected to an amplifier constitutes no more than 60 percent of the rated output of that amplifier. Loudspeakers should be high impedance tapped at 70 or 100V, with power tapped at no less those 15 watts. Circuit the loudspeakers so that speakers above the stage position may be switched off to increase overall room gain before feedback. The use of 8-ohm speakers shall be limited to areas where high SPL's are needed.

The system must provide for multiple microphone and line level inputs from each room as well as providing for at least one-line level return to each room for the purpose of in⅓room session recording. The Digital Signal Processor (DSP) input configuration on the processing devices shall provide for every room input to be actively connected without need for physical patching. DSP devices should be interconnected globally utilizing the CobraNet standard for transmitting audio and control signals via Ethernet connections. The system shall also provide for the ability to virtually route audio signals anywhere in the facility which will allow for routing of multiple BGM sources to their corresponding outputs.

The ballroom A/V system shall cater for the following:

Projection and large screen display of high-speed Internet access and laptop computer screens, video images from DVD players' digital visualizers, electronic document camera, video conferencing, including all necessary auxiliary inputs, audio support for all the above listed plus a multi-changer CD, and lectern, microphone system as described in section 12.4. Background music system; Image monitoring at the lectern location; Touch Panel Remote Control system integrating all of the above plus electric screens, projector hoist systems, house lighting system, voice evacuation and fire alarm system, and electrically operated door closers, blinds or drapes. An iPad application shall enable one to operate the complete Control System remotely over a wireless network.

The system must be designed to accommodate the number of subdivisions of the main ballroom space which may entail full provision of services described to each subdivision, the essence being that each sub divided space can stand alone or be combined. The projectors must retract into the ceiling void when not in use and the screens, if permanent, must also retract.

The ballroom should be fed by an integrated mixer/amplifier system having a minimum of 4 microphones per 300 m^2. Divisible rooms should have the means to connect all rooms or operate separately.

The sound system electronics racks shall be in the back of house. The rack shall be positioned so that it can be serviced from the front and the rear without being moved. The ballroom sound system should be capable of producing 95 dBA at 125 cm above the floor.

14.1.7 Food & Beverage Outlets

All Restaurant space audio systems shall be capable of producing 85 dB SPL at 170 cm above floor for standing persons. The frequency response should be ± 7 dB from 100 HZ to 7000 Z. The system shall be free from audible distortion. The area shall be covered by distributed loudspeakers serviced by one or more independent power amplifiers Distributed speakers shall be mounted in the accessible areas of the ceiling wherever possible. Spacing should be arranged to provide even coverage throughout the audience area, while maintaining an aesthetic or unobtrusive relationship with the ceiling architecture and other elements.

Coordinate the location of loudspeaker devices with the acoustical environment. Layout at partitioned spaces shall accommodate the various room configuration without disruption of the sound coverage, when the partitions are removed or replaced.

The volume control for the Restaurant shall be located at the central sound system equipment rack or in an adjacent non-public location.

14.1.8 Staff Training Room

The training room shall be able to comfortably accommodate up to approximately 30 persons, in terms of capacity. The room should be accessible, and designed with wheelchair users in mind.The room shall be equipped with the following A/V equipment:

- An LCD monitor, or a ceiling recessed tensioned electric screen or a pulldown screen shall be provided. The largest screen possible shall be provided at a minimum height off the floor of approximately 1.2m to the bottom of the screen
- The maximum viewing distance shall be no more than 4 times the screen height. The minimum viewing distance shall be 1.5 times the screen height. The viewing angle of the screen shall be no less than 60° visibility being critical consideration shall be made for different aspect ratios, some of it in 4 x 3, some of it 16 x 9, to accommodate the needs of a variety of source content, which can be anything from a website, a PowerPoint presentation or DVD content.
- The projectors shall have two VGA inputs for a PC or laptop, as well as a component and/or composite input for a DVD or camera feed.
- The projector shall be mounted on a Smart-Lift Automated Projector Mount, and recessed in the ceiling space. The operating system shall be control options of a low voltage controller, and internal and external wiring options, and accepts 5-30V AC/DC for triggering commands

- Provide a teaching desk equipped with desk microphone and a desk light as well as a pull-out tray to house the PC keyboard, hand held microphones and
- A control system that will control lighting, screen, projector etc., by effectively allowing communication between equipment. A touch screen user interface shall be provided.

14.1.9 Video Conferencing System

The Videoconferencing System shall have the facility of easy-to-integrate audio, video, and data communications across IP-based networks, including the Internet. All equipment and components shall comply to H.323 signaling protocol, where multimedia products and applications from multiple vendors can interoperate.

The system shall allow 6 to 8 participants to comfortably communicate via Videoconferencing in a point-to-point call, yet as standard, will enable multiple locations to communicate with each other by connecting to an external Multiport Control Unit (MCU). The MCUs shall be equipped with a software package that could schedule a conference with attendees in different time zones.

Dual Video Streams shall be available to allow "presentation" (content) audio-video stream to be created in parallel to the primary "live" audio-video stream. This second stream will be used to share any type of content: slides, spreadsheets, video clips etc.

The integration of teleconferencing (video over IP) equipment with A/V presentation systems that will enable cameras strategically placed in the room to be synchronized to individual microphone activation, and microphones of the boundary layer type strategically located to pick up voice from each participant at the table.

The presentation material could be served up either via a networked computer or a media player. With increasing interest in HD video and larger files of presentation content, the computer or media player should include more robust video and audio processing and output capabilities.

The software of the conferencing systems should allow for multiple view options from all cameras to and from the remote locations. Meeting rooms with these A/V presentations and conferencing systems must be integrated with touch-panel systems controllers that also ideally incorporate the lighting control system, and other advanced functions.

The designer shall thoroughly investigate the interworking between protocols H.323 and SIP (Session Initiation Protocol), and adopt the most suitable and cost-effective solution for the facility location.

The office space planning approach called "fluid architecture" using a design approach where there are less rigidly cordoned off cubicles, more unassigned spaces for telecommuters to "park" on days when they are physically at the headquarters, and more open spaces for groups to convene for planned or spontaneous collaboration. Fluid architecture provides setting for the "task team" trend. This approach should be enabled by wireless networking. Software that enables group collaboration from remote locations may be required by the tenant. With these programs, participants could see a shared desktop on their screens, or they can also use the A/V system in meeting rooms.

14.2.1 Digital signage

Digital signage is used for wayfinding, advertising, and merchandising. Dedicated media players should drive the repeating content in these applications using flat panel displays.

14.3.1 Networked A/V Security

Since A/V systems are being migrated onto enterprise data networks, the A/V system should maintain a security posture in alignment with their security goals. Unauthorized access to the A/V system could impact the three principal goals of security — confidentiality, integrity, and availability. Additionally, security flaws within IT-attached AV devices could potentially provide a platform from which to launch attacks at other IT systems.

The designer should develop security requirements in the initial planning and design phase of the A/V project. The security requirements should be included in any bid documents and added to the Statement of Work (SOW). If the security requirements cannot be determined prior to the integration contract award, language should be put in the tender that requires the tenderers to specify standard security configurations and that specific security requirements will be determined after award with equitable adjustments for changes.

In all cases security requirements, should be agreed upon and added to the statement of work, project plan, and acceptance criteria prior to the final design signoff with the responsibilities and deliverables of each party, including the network configurations furnished by the network provider.

14.3.2 AV / IT Security Framework

As guidance, the steps in the process of determining the project specific security requirements are as follows. These may be altered/modified to suit the individual project:

- Phase 1 - Determine the Organizational Requirements
  - o Verify that you have input from all the stakeholders
    - — A/V Integrator
    - — A/V Consultant
    - — Owner
    - — Operational Security
    - — Information Security
    - — Physical Security
- Phase 2 - Analyze the Security Risk
  - o Create a threat model
  - o Implement a risk register

- Phase 3 - Establish a Risk Response
    - Establish a risk response
    - Agree on a mitigation plan
- Phase 4 - Documentation

Typical Security Documentation Includes:

- Hardware Software inventory
- Ports, Protocols, and services
- OS and package versions
- Network topology
- High Level Network Diagram
- Logical Topology
- VLAN assignments
- Connection Points
- Traffic Flow
- Access Control Lists
- Roles and permissions
- Password Policies
- Physical Topology

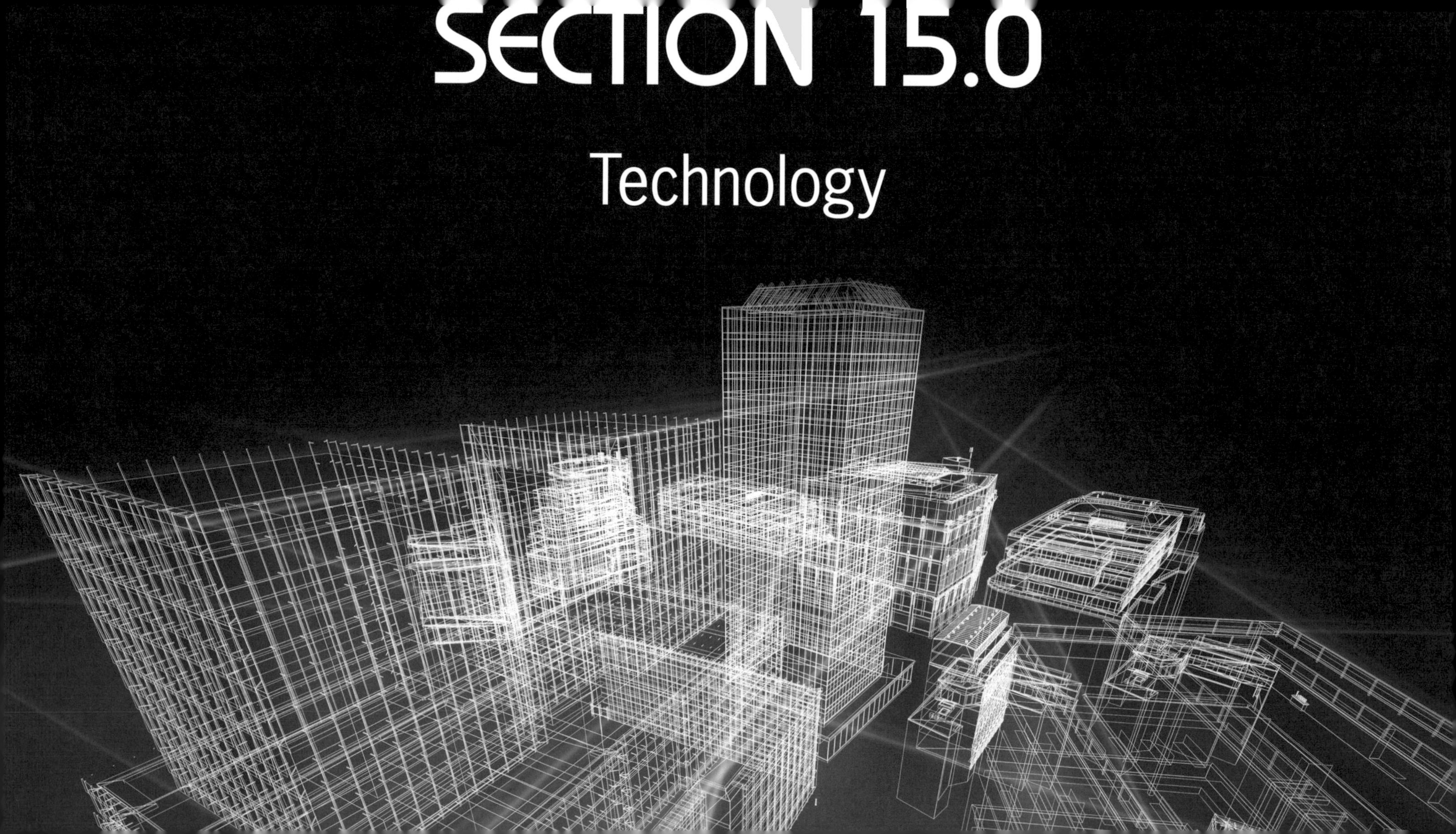
SECTION 15.0
Technology

15.1.1 Technology Statement of Direction

For the owner's brand to compete in today's ever- changing environment, technological standardization and integration of systems is essential. We view technology as a competitive tool that contributes to the improvement of customer services and operating efficiencies. As the importance of access to information increases, so do customer demands for such services. Through technology, The building occupants expect the building management to react immediately to their increasing demands.

To efficiently implement new technologies, the Owner need to be positioned to respond to change. To capitalize on competitive advantages the Owner must take advantage of opportunities to implement new and innovative services in a timely manner. Brand standardization provides the most efficient and cost-effective means to respond.

Through standardization across the company, resources can be focused towards the continued development of a technology platform, and the negotiation of pricing and brand implementation can take place country wide and regionally. As new innovative ideas are introduced, the Owner can work quickly and efficiently through the chosen strategic providers to install new applications and services throughout the country and further afield. Strategic partnering with global and regional industry leaders who are committed to the hospitality industry will ensure that the building management team can continue to provide innovative applications and services to customers.

Prior to any technology commitments being made, the Owner must be consulted for their input, advice and approval of any proposed systems, solutions, vendors, resellers and distributors.

15.1.2 Physical Environment

General

This document is a guideline to the professional engaged in Information Technology (IT) design disciplines. It is to be used in the preparation of detailed design and specification documents for construction of mixed-use real estate projects. The intent of this document is to establish minimum and sufficient technical criteria, and typical functional requirements or "standards" for each branded real estate project.

These "standards" are not to be considered absolute; special consideration will arise in each different facility. In addition, the fundamental technological dependency of the systems design afforded by technical advances may suggest alternative approaches.

Meeting the standards below for the physical environment of the Data Centre is necessary to protect and maintain critical property systems. These standards encompass location, security, electrical, room layout, disaster prevention and environmental requirements.

- The Data Centre is the central distribution point for all voice, data and TV systems
- The Data Centre houses all vital systems including rack mounted business systems, Local Area Network (LAN) Backbone, Wide Area Network (WAN), iPBX, related equipment, UPS, MATV and all other peripheral business systems.

- The Data Centre should preferably be located near to the majority of system users and main riser/s to reduce wiring costs by minimizing cable runs
- Due to potential risks from physical damage, water ingress, explosion, vibration and HVAC loads the Data Centre must not be located adjacent to nor utilize an exterior wall of the facility.
- Water, steam, sewerage pipes or HVAC ducts and pipes must not pass through or over the Data Centre. If this restriction is impossible computer equipment must not be located under these ducts and pipes. The pipes shall be installed with a protective metal shield and the Data Centre must have waterproof covers readily available in the event of a leak or rupture.

15.1.3 Security

Due to the large investment in the facility's voice and data systems, enforcement of a strict security policy for the Data Centre is essential.

- Due to potential risks from physical damage, water ingress and explosion the Data Centre must not be located adjacent to nor utilize an exterior wall of the facility.
- All entry points to the Data Centre must have an electronic locking system preventing unauthorized access. Restricting Data Centre access to security screened authorised personnel is essential.
- The preference is to use two-factor authentication. Biometric identification is becoming standard for access to sensitive areas of data centers, with hand geometry or fingerprint scanners considered less invasive than retinal scanning. In other areas use the less-expensive electronic access cards.
- All access points should be monitored by CCTV as should the interior of the main computer room. There must be no obstructions to block surveillance by CCTV cameras
- There should be signs on all doors marking the rooms as "restricted access" and prohibiting food, drink and smoking in the Data Centre

15.1.4 Electrical

Power for the data room should consist of a UPS and power distribution. UPS systems for this application should be typically line-interactive for loads up to 5 KVA, and double-conversion for loads above 5KVA. UPS systems greater than approximately 6kVA are typically hardwired from an electrical panel. Electrical power supplied to the UPS should be from a dedicated branch circuit from the Main Distribution Board, used by the data center only.

- In general, doubling the required amount of power for the Data Centre will provide for any future demands. Power receptacles need to provide power for both 20A and 30A devices.
- Every outlet in the Data Centre is required to be on emergency power. To avoid power issues, main circuit breakers should be used for the Data Centre and should not be shared with other locations within the facility

- In addition to an Uninterruptible Power Supply (UPS) the Data Centre should be connected to the facility's back-up emergency generators
- Provide an adequately sized UPS System to protect the equipment from under-voltage, undercurrent, voltage surges, and current surges with the following features:

(a) Supply battery power until either a successful shutdown or the emergency generator can supply power to the equipment.

(b) For power cable management, all IT equipment should be plugged into a rack power distribution unit, which is plugged into the UPS.

(c) Provide an integrated network management web card to allow critical remote UPS monitoring such as low battery, bad battery, on battery, overload, low runtime, etc.

(d) The UPS system should be hardwired from a separate dedicated electrical feed and circuit breaker

(e) The branch circuit should have an insulated wire conductor, equal to the size of the phase conductor, for grounding the equipment. The branch circuit grounding wire should be tied to a single insulated ground point at the distribution panel, and a single grounding wire should run from the distribution panel insulated ground point to a service ground or suitable building ground. This should be a dedicated ground, not neutral. Conduit or the utility box itself should not be used as the grounding means

(f) This clean power SHOULD apply to all equipment in the Data Centre.

(g) Provide a dry contact I/O sensor which should notify administration/security, when the server room door is opened.

(h) Provide a means of manually disconnecting the power to all equipment in the Computer room. This unit should be located adjacent to the exit door of the Data Centre or IT Manager's office, but not accessible to unauthorized employees. The power cut-off switch should be achieved by providing a shunt-type circuit breaker as a main in the panel serving all equipment in the Data Centre

15.1.5 Disaster Prevention

The type of fire protection systems selected should be tailored to the specific needs of the facility. A professional engineer/fire protection engineer must design all fire protection systems for the Data Centre and riser closets

- The Data Centre must be provided with smoke detectors connected to the main fire alarm panel and be fully protected with a water based fire sprinkler system Where the Data Centre, PBX room (or combination of the two) has a false ceiling (dropped ceiling) and/or a raised floor, smoke detectors should be placed in the space above the false (dropped) ceiling, in the room itself and also under the raised floor
- All smoke detectors should be connected to the main fire alarm control panel and must enunciate at a minimum in the System's Managers Office, the Data Centre, the Telephone Operators and the Central Alarm Panel
- The Data Centre shall have an automatic pre-action dry sprinkler system that shall be independently zoned for the Data Centre, equipped with an audible and visual alarm located at the main fire control panel. The sprinkler system shall feature automatic discharge via cross-zoned smoke detectors, timed (30 seconds) delayed discharge on the second alarm
- Provide the facility for a manual shut-off switch (where allowed by local law), a manual discharge mechanism, and shutdown of power to the cooling and ventilation devices upon sprinkler system activation

- In addition to the pre-action dry sprinkler system, provide at least three (3) portable or other extinguishers appropriate for electrical fires. One adjacent to the entrance, one adjacent to the computer equipment and one in the UPS room. A sign shall be located adjacent to each portable extinguisher indicating the type of fire for which it is intended.
- The use of total flooding CO_2 or Halon systems is not approved unless demanded by the local authority having jurisdiction
- The Data Centre must have a drain to prevent flooding should a pipe break and to manage water run-off from activated sprinklers. Plastic sheets large enough to cover all computer equipment should reside in the Data Centre
- A manual Pull Station must be located outside the main door to the data Centre as part of the facility's fire detection and alarm system
- System media stored in the Data Centre must be housed in a separate fire rated storage cabinet
- The daily back-up media of every critical system must be stored in a fire proof safe which is remotely located from the Data Center
- A NO SMOKING, NO FOOD, NO DRINK policy must be strictly enforced in the Data Centre
- Detailed safety, security and emergency instructions must be posted in the Data Centre

15.1.6 Environmental

Cooling and Ventilation

Monitoring humidity is essential. Too much humidity causes corrosion and too little may cause equipment damage due to static electricity. The Data Centre should be maintained between 21⅓C and 23⅓C, and Relative Humidity between 40% - 50%.

- The Data Centre should be equipped with an independent, dedicated process cooler (precision air conditioning) which is properly rated for the allocated space. There should also be a second air conditioner which is capable of supporting the full load of the room. The secondary system may be an independent unit or a portable unit. The primary and secondary units should be wired to the emergency power generator and the emergency Power Off Switch
- With a raised floor environment (minimum height 60 cm) the primary air conditioner should be a down flow unit discharging to the raised floor plenum. The raised floor should vent the air through perforated panels. The back-up air conditioner should be able to meet the full design cooling capacity of the Data Centre
- The Data Centre should be equipped with a high temperature alarm and at least two water alarms (under the raised floor) which enunciate at a minimum in the Systems Managers Office, the Data Centre, Fire Panel, Manager on Duty and the Telephone Operators
- The air-conditioning unit/s should be on the facility emergency power source
- Utilize the facilities available with the UPS system integrated network management web card to provide environmental monitoring, temperature and humidity
- Ventilation for the battery back-up room should be provided in accordance with section 502.3 of the International Mechanical Code, to prevent a dangerous accumulation of flammable gases. The battery

room ventilation system should be designed such that the room hydrogen concentration is limited to no more than 1%

- Air conditioning should be adequate based upon the equipment housed in the Data Centre. To determine the air conditioning requirements, record all equipment KW doubling the Total KW. always verify that the environmental conditions meet or exceed the equipment specifications. The following table lists the general equipment you will find in the Data Center including the power requirements and KW output.

Lighting

See Engineering Design Criteria

15.1.0 Computer Room (Data Centre)

The Data Centre represents one of the most sensitive areas of a facility for fire prevention, water infiltration protection, loss control, safety and security. This guide is intended to address the planning, design, and construction of these facilities.

I. Space Planning Charachteristics

The Data Centre must house all critical computer processors, workspace, storage space and be "sized" accordingly

(a) **The Data Centre complex consists of:**

- Computer room housing all main business system servers and iPBX
- Uninterruptable power supply room
- iPBX battery room
- iPTV/ video on demand head end room
- Systems Manager office
- Repairs and storage equipment room

(b) Space allocation for the Data Centre Complex should be carefully studied

(c) The computer room must house the telephone system (iPBX), MDF and all ancillary processors such as call accounting and voice mail. This should be clustered in one area of the room adhering to the manufacturer's specification for system footprint and access. The area allocated to the telephone system must also allow space for one additional cabinet should expansion be required

(d) The iPBX battery backup should be in a separate room

(e) The computer room must house all rack servers required for all other business system

(f) Computer racks must have seismic bracing and proper grounding

(g) Computer racks should have adequate clearance per manufacturers' specifications but be no less than 1.25m front and 1.0m rear clearance

(h) Due to the air conditioning load of the UPS and the subsequent rapid rate of rise in temperature should air conditioning fail, the UPS must be in a separate room

(i) The main day to day entrance to the computer room should be via the system managers' office to increase security

(j) From raised plenum floor the room should have a minimum height of 3.0m to the ceiling

(k) The IT Managers office should comprise of two workstation positions. One for manager and one for technician

(l) The Repairs and Storage room must have work counters and media storage cabinets along with a floor mounted fireproof data safe

(m) The Computer Room must have a workstation position along with a work bench for use by systems technicians and security cleared external maintenance technicians

II. General Design

(a) The perimeter of the Data Centre will be constructed with a two (2) hour rated fire wall. Walls will run true floor to true ceiling and all penetrations will be sealed. Walls to be painted or protected with a non-combustible finish containing no exposed cellular plastics

(b) Fire rated door assemblies shall conform to the test requirements of 2012 International Building code (IBC) section 76.5.1 or 716.5.2 and fire protection rating in table 716.5

(c) Walls, ceiling and doors should be isolated from other occupied areas

(d) The Data Centre should be equipped with a "raised floor" for cable tidiness, cooling and flood control. The floor must be structurally strong enough to support the weight of fully loaded server racks, the telephone system and transportation devices for movement of equipment

(e) For safe equipment delivery and removal, the raised floor must be accessed via a ramp. This should be accessed via the main double door entrance

(f) Floors must be fire retardant and have a flame-spread rating of 25 or less, anti-static sealant and be equipped with a gravity feed drain with back flow prevention. If a drain is impossible or impractical the facility must maintain one portable submersible pump in the event the Data Centre becomes flooded and must have access to the electrical supply from the facility's emergency generator

(g) Antistatic floor finishing (no wax) is recommended for raised floor tiles or sheet vinyl

(h) Unless restricted by local code, there shall be a self-closing, self- locking double width door set to the Data Centre with a minimum fire rating of 90 minutes for the complete door assembly. One of these doors must have a wired glass or ceramic glass window measuring no greater than 50cm square

(i) All Data Centre doors accessible from the access corridor must be equipped with a digital programmable door lock or electronic lock from the main facility e-lock system

(j) All internal walls and partitions shall have a one-hour minimum fire resistance rating and doors shall have a minimum 30 minutes fireproof rating

15.2.1 PBX Room

Traditionally facilities have provisioned a separate PBX room. Since the advent of internet protocol communications, voice and data are composite to the structured cabling enterprise and are combined in one computer room and therefor this separate room is no longer required

15.3.1 IP TV/ Head End Room

The room is designed to accommodate all of the equipment required to deliver interactive TV and entertainment services to the public areas and residential apartments. An additional access door is made available for authorized third-party contractors, in the event that the systems require attention, maintenance or rebooting out of hours. Standards for UPS support and air conditioning apply as with the main Data Centre.

15.4.1 High Speed Internet Access

Providing reliable high-speed internet access (HSIA) and WiFi to occupants and the facility operation, and making the building future ready with high bandwidth requirements is critical.

Whether delivered by fixed wire or WiFi this service should be treated as an unfailing "utility". Basic component parts of the HSIA shall consist of the Fiber line, Firewall, Routers and Active Equipment. Additional Active Equipment shall be distributed throughout the facility via fiber cabling. Very early in the construction project, the HSIA Vendor should work with Owner's IT department and the solutions partners to design an elegant plan that took the building layout and connectivity needs into consideration equally.

I. Internet Management Platforms and bandwidth requirements should be determined in consultation and agreement with the Owner's IT department.

II. Bandwidth capacity requirements are based upon the property needs and corporate mandates. Factors need to be taken into account such as;

(a) Average number of devices per user e.g. 2 and rising

(b) Utilization rates of average user e.g. 60%

(c) Types and volume of user internet traffic are determined by business mix, meeting spaces, conferences, trade shows, movie downloads and wedding celebration parties

(d) Administratively consider the impact of on line booking channels, cloud based applications, emails, social media, web browsing, file downloads and VPN connections.

(e) Average circuit utilization might be 45% over a 24-hour period but peak utilization at times could be 80%

III. Bandwidth solutions from ISP's and Telco's should be evaluated for flexibility, scalability, cost, reliability, burstability and robustness. Dependent upon the geographic location it may be advisable to obtain circuits from multiple providers for resilience

IV. Bandwidth management technology should be implemented. With an ability to cap bandwidth and the potential to set bandwidth tiered pricing so that network resources can be managed cost effectively. This can especially assist where a user is downloading large multiple files hogging bandwidth from other users

V. Tiered bandwidth management enables free basic access with optional upgrade pricing for higher bandwidth. This is especially useful for conference pricing when there is a significant uplift in bandwidth demand

15.5.1 Main Distribution Frame

The Main Distribution Frame will be a purpose built, located in a secure room within the building that will act as the "Demarcation Room" for all telephone and data services. The room may also be the Main Equipment Room (MER)/Data Centre, except where there is a specific requirement for the demarcation room to be physically separate.

I. Conduits will enable egress and exit of services from the building via this room, with additional provision allowed for future services.

II. Conduits will be located such that the distance to the service provider's delivery point on the boundary of the property are kept to a minimum.

III. Data services in the form of fiber leased lines and copper digital services shall terminate in this room.

IV. Termination will be to a demarcation cabinet or within a provided data cabinet, directly interfacing with the Local Area Network (LAN).

V. Links will exist between the MDF and the MER/ Data Centre, extending the delivered service to the distribution infrastructure. These links will be multimode fiber/ Cat6a for data services, and multi core Cat3 for analogue telephony.

VI. UPS provisioning should be included in the building support system or if physically difficult must have a self-contained system capable of supporting any active equipment in the room.

VII. The room shall be physically protected, and access monitored bysecurity systems

15.6.1 Uninterruptible Power Supply

A suitably sized "in-line" UPS for the entire Data Centre should protect all equipment. Individual UPS's for each device is not acceptable.

In locations where the local grid power supply is unstable it may be advisable to have two identical UPS's available for resilience

I. All computer devices in the Data Centre will receive their power from a properly sized UPS with a minimum 15-minute sealed battery back-up supply. The UPS should include a static by-pass switch that enables an uninterruptable cut over to utility power as well as a disconnect switch. Telephone Systems are usually equipped with their own battery supply and need not be wired to the UPS if so equipped but should be connected to the facility's emergency generator

II. The Data Centre and all administration, front office, point of sales computer devices should be on a "clean power" electrical supply sourced from the main facility power distribution board

III. All computer power outlets should be of an alternative pin type to those for general purpose use to protect against unauthorized appliance use. Failing this they should be color coded and marked "for computer use only". The vendors of computer systems should be notified before ordering of the pin type and voltage being used for computer systems

IV. Where local power conditions culminate in frequent and/ or prolonged power outages it is necessary to place (at a minimum) key point of sales workstations, front desk workstations, reservations, telephone operators and security on UPS supply. Related printers should also be included. Additional workstations may also be advised as determined by the operation and facilities of the facility. This will enable the facility operation to remain efficient providing expected service levels to users

V. The facility's emergency generator should be wired to the main UPS and have sufficient capacity to service it in the event of a full power failure

VI. The UPS will be housed in a separate room adjacent to the Data Centre. The UPS room should adhere to the manufacturer's specifications for environmental conditioning. The UPS should be capable of transmitting real-time information to the Building Management System.

VII. The distribution panels for the UPS and utility power should be located in the Data Centre

VIII. The Data Centre should be equipped with a "secure" Emergency Power Off switch located outside the Data Centre and adjacent to the door. This EPO switch will cut all power to the Data Centre in the event of an emergency

IX. The Data Centre should be equipped with a battery powered emergency light. all primary lighting will be wired to the emergency power generator

X. All power outlets in the Data Centre should have isolated grounds. Under no circumstances should the junction box, conduit, water pipes etc be used to ground an outlet which supports a computer system or its peripherals

15.6.2 To determine the size of UPS required;

I.List all of the equipment that requires protecting. Include servers, racks, hubs, routers, switches, a/c controllers, monitors, terminals, external hard drives, user workstations in critical areas, key POS devices, MDF's, IDF's, WiFi access points and any other critical equipment.

II. Determine "load" by calculating the VA rating. VA Ratings will help determine the size of UPS required.To find VA (VOLT-AMPS), find the Voltage and Amperage requirements for each piece of equipment. Multiply these numbers together: VA = Volts x Amperes

III. If components are measured in Watts, then multiply the number of Watts by 1.4: VA = WATTS x 1.4

IV. Total the VA Requirements by adding the VAs of each piece of equipment to find the total VA.

V. It is necessary to provide at least a 15-minute UPS for an in-line generator in order to transition to generator power. Provide a 30-minute UPS for facilities where there is no backup generator. It is recommended to oversize the UPS to take into consideration any future expansion needs.

15.7.1 Cabling System Standards

(a) **General**

The building shall be equipped with a distribution network of cable trays and baskets. The containment shall be located within communications risers, ceiling channels and raised floor voids. Separate containment should be installed for voice/ data and power cabling.

(b) **Design Over view**

Rapid technological changes, increased cost and deregulation make it necessary to view the building cabling as a strategic investment. In building distribution networks must meet both the near and the long-term service needs.

Traditionally Information and communications technology systems have focused on designing structured cabling and technology systems only for Information Technology (IT) groups. With audio visual equipment, cellular enhancements, lighting and mechanical systems coming to rely on structured cabling systems (SCS) for connectivity, designers must fully understand each section needs and develop unified structured cabling design solutions that meets the demands of multiple clients occupying the facility. The SCS must be fully certified with a relevant 20 Year Performance Warranty.

Blown fiber network infrastructure technology should be considered to accommodate the continuous bandwidth- hungry applications and ever-increasing network speeds within a constrained budget.

Air-jet technology will provide benefits for high-density, high-speed enterprise networks including:

- Virtually unlimited fiber pathway and bandwidth capacity
- Immediately scalable, real-time already future proofed sustainable network
- Fast and easy fiber installations, upgrades, and network moves/ adds/changes
- Continuous fiber runs, eliminating splicing and potential points of network failure

The designer is required to provide a global, vendor independent cabling plan that delivers performance and flexibility at a reasonable cost.

The structured cabling system shall be designed as an all-encompassing end-to-end solution based on a hierarchical topology.

15.7.2 Cabling Infrastructure

In order to meet the total, combined and separate objectives of the IT facilities and more, the concept of systems separation shall be fully applied.

(a) **Physical Separation**

Security policies for change control management may require moving facilities managed systems to separate rooms from the networking equipment. This allows easier access by technicians or electricians when maintaining facilities related equipment. The designers must work with each group to determine the level of physical separation required.

(b) **Visual Separation**

Visual separation of systems should be achieved using different color wires, within a shared pathway or space. Visual identification of systems is essential in a congested ceiling environment or within a crowded telecommunication room.

(c) **Pathway Separation**

For businesses that service different clients or types of businesses subject to varying regulations, separating the cable pathways may be necessary. This information must be disclosed during the pre-design phase.

15.8.1 Intermediate Distribution Frames

Intermediate Distribution Frames are purpose made closet locations for cross connecting data provisioning to a specified area (e.g. an entire floor, a connected building or conference area), from the MDF and the network. The following standards should be followed: Horizontal Cat6 cabling from the IDF to the endpoints will be no longer than 90m. Total distance from IDF cross connect point, to endpoint shall be no more than 100m in total. Patch leads for cross connecting end points to the IDF switching shall be colour coded to identify data and telecoms ports. Horizontal cabling shall terminate in the IDF into patch panels in the IDF enclosure. Cross connect cables (patch leads) will connect patch ports to the relevant network or telecoms ports on network equipment locations shall be kept locked and secured to deny unauthorised access and avoid interference. The IDF shall be large enough to house an appropriately sized data cabinet for the location. This location shall be climate controlled by the immediate environment or from a purpose built environmental management system. All network based systems relevant to the areas served by the IDF shall be connected to the IDF switching infrastructure. Data for each system shall be isolated from other systems by the use of VLANs. An elaborate VLAN structure will exist on the network to accommodate all data systems (see 13.9.7) IDF equipment will be monitored similarly to the rest of the network with network management tools (see 13.9.7) Physical security of the IDF shall be maintained at all times, and access will be monitored.

15.9.1 Local Area Network Standards

General

The designer shall plan to setup a scalable and futuristic network infrastructure, which will not only serve today's needs, but also future proof the network for voice, video and e-commerce services. The basic principles of the network shall be reliability, scalability and simple to manage. The entire product line should have a consistent architecture to reduce total cost of the network ownership for on-going support, maintenance and training. The following standards should be followed:

I. The LAN will be designed to separate all network based services from one another with VLANs.

II. Equipment requirements will depend on the number of ports required. The number of outlets, powered and passive in each section of the building will be taken into account, with a minimum 10% allowance for growth.

III. The required LANs will be identified during the network design phase taking into account the following;

(a) Access to any LAN (VLAN) will be limited to authorised users only, and only to systems the user is authorised to use

(b) Security shall be designed into the network configuration to deny unauthorised access, accidental or malicious.

(c) The configuration shall be detailed and documented. All configuration files shall be backed up after every change. All changes shall be registered and approved.

(d) There should be POE ports for Ethernet powered devices such as IP phones and CCTV cameras

(e) Internet access (firewalling) shall only apply to VLANs where this is essential and will be managed via proxy server and firewall. CCTV and Access control VLANs will NOT be open to the internet under any circumstance

(f) QOS to prioritise voice and video traffic

(g) SNMP traffic monitoring will be enabled on all devices. SNMP v2c shall be used as a minimum. Complex community string name shall be chosen. Default string 'public' shall be removed from device as a policy.

(h) There should be alternative routing between the MDF and IDF's (redundancy, STP). Redundant links shall be installed where possible, with routing protocols employed to manage active fail over

(i) Network connected environmental monitors will be employed in all IDF to alert changes in temp and humidity

(j) Alerting system linked to monitoring Secure remote access via VPN for support services so that restricted access only provided to the supported service

(k) Secure remote access via VPN for remote workers with restricted access only to authorised applications and data.

IV. Switching technology in the MDF and IDF will support the following;

(a) Full range of VLANs (i.e. Extended range, 1-4094)

(b) Port management

(c) Access restriction

(d) SSH remote administration access

(e) All switching will be sourced from the same manufacturer

(f) Extended warranties should be taken out to secure the devices

(g) Access to same day replacement or spare devices on site

(h) Firmware upgrade remotely from tftp server

(i) Configuration backup centrally to secure tftp server

(j) Automated VLAN table propagation (e.g. Cisco VTP)

(k) Mix of powered and non-powered ports depending on requirement

(l) MDF core devices to support 40 Gb throughput

(m) Ports not in use will remain in a state of shutdown until required

15.9.2 Core Layer

General

Starting in the "core" of the network, this switch shall be a modular switch and should have features and port densities that will not only serve the current topology, but will allow the topology to mitigate and expand into larger and more sophisticated network.

As the network is very critical to the facility operation and no downtime is acceptable, designers shall consider various reliability/redundancy options like dual core switch configuration.

I. High data transfer rate required

II. Load sharing configuration should be used to maintain high speed data transfer rates

III. Specification should be for high-speed low latency devices

IV. Multiple data paths will be employed to ensure high availability

V. Device (and vendor) should be selected on cost, supported bandwidth, port density, features and manageability. Requirements and port count needed will determine a chassis or stack switch configuration

15.9.3 Distribution Layer

Routing and policy based network connectivity will be employed at the distribution layer

I. Quality of Service (QoS) will be applied at this level

II. Access layer switches will aggregate at this point

III. This layer will serve as an aggregation point for access layer switches

15.9.4 Network Availability

The network shall be designed for optimal uptime, making use of redundancy features of all network devices. For example, redundant power supplies, network interface cards, trunks and routing protocols.

I. Advanced network monitoring tools shall be deployed to monitor key counters on all equipment, alerting when status changes occur that could lead to a network failure

II. Service Level Agreements (SLA's) shall be required, to guarantee that service levels are met

III. Preferred network devices will incorporate ISSU (In Service Software Upgrades) where updates can be applied without the need for restarting equipment and introducing downtime

IV. Physical, operational and infrastructure level security procedures will be observed to reduce the risk of malicious interference

V. Manufacturer warranties, on-site spares of key devices and comprehensive configuration logs will all be employed to reduce the effect of a faulty or damaged network device

15.9.5 Network Security

The physical security of network devices in the mechanical and electrical room and termination end points will be observed.

I.Devices will be installed in locked enclosures. Keys will be stored securely, and access logged.

II. A Strong Password policy to be applied to all network devices, servers and systems. Passwords will be changed periodically in line with Domain network policy

III. Terminal Access Controller Access-Control System TACACS+ server will have employed to manage user authentication, authorisation and accounting on switching and routing devices

IV. Firewall device will be configured to block unauthorised network access and to configure remote access VPN. All but essential ports will be configured “blocked” by default and only opened by authorised personnel

V. Administrative internet connection shall be physically separate from the guest High Speed Internet Access (HSIA) in guest rooms, meeting rooms and business centre

15.9.6 Firewall, Antivirus, Malware

It is essential that steps are taken to protect customer data and ensure that strong security measures are in place to protect the facility network, payment systems and Wi-Fi services. It is critical to keep all firewalls, anti-virus and anti-malware software up to date with the latest versions and definitions. Companies that specialise in these software's monitor threats constantly and regularly update preventive measures and improvements to thwart attacks. Additionally, the IT department updates all operating system security updates and patches when notified.

I. The policy of the Owner is as follows; All of the facility's data circuits and internet connections are protected by a robust and adaptable firewall

II. End users will not be able to make changes to, or turn off, anti-virus or the firewall. This will be managed by profiles configured in the AV management centre

III. Alerts will be sent to the network administrator when virus or malware vulnerabilities are detected

IV. Updates will be scheduled to match the update release rate of the software vendors

V. In addition to blocking unwanted inbound traffic it is good practice to selectively block outgoing traffic. Many breaches involve malware that becomes resident on the network and then tries to send sensitive data to hacker systems via the Internet. Secure data must not leave the network without the network administrators' knowledge

VI. Protection of on premise Wi-Fi because devices are more connected to the Internet. Users expect to have wireless communication whilst in the facility and wireless networks can potentially expose sensitive data especially in a facility environment

VII. Use of two factor authentication when permitting remote access to an internal network as it is essential that this access is restricted and secure. At a minimum, access is only granted to individual (not shared) user accounts using two factor authentication and strong credentials. Remote access will also be logged so that an audit trail is available.

15.9.7 Network Management

The designer shall propose a cost efficient, comprehensive Web-based graphical Network Management System.

Network devices should be monitored with industry leading network management software.

I. Sensors will be deployed, for example to monitor vital device counters, network traffic, status changes, data throughput and temperature change

II. Sensor tolerances will be configured to provide advance notice of tolerance breaches e.g. CPU and memory usage, temperature change, abnormal traffic patterns, port status changes etc

III. Physical security of all network distribution devices shall be observed. Devices will be installed in adequate data cabinets. Cabinets will be labelled and locked, and access logs kept.

IV. A software and firmware “patch and upgrade” schedule shall be observed, to maintain all devices

15.9.8 Hardware

I. Network hardware selected for installation shall be from a single manufacturer and will be of Enterprise grade construction, capable of high levels of transaction performance and data security. The hardware will be backed by robust manufacturer warranties extended or renewable to cover the expected useful life of the device.

II. The devices will of suitable data throughput to more that cope with expected data transfer levels.

III. The devices will be purchased with the knowledge that the manufacturer development roadmap includes support for the chosen devices for the expected working life of the device i.e. devices not close to end of life and manufacturer support.

IV. Devices will be chosen with built in redundancy for vital components such as PSU and NIC.

V. Upgrades either to hardware, Firmware or both will extend the useful life of the device or enable added features.

VI. Devices will support In Service Software Upgrades (ISSU) to reduce downtime for firmware upgrades.

VII. Devices will be installed to the manufacturer’s recommendations to avoid affecting warranties. All environmental recommendations will be observed.

15.9.9 Switches, Routers, Peripherals

Switching, routing and network devices will be sourced from a single industry leading manufacturer. (Examples of which would be Cisco or HP). The devices will be installed and configured to support manufacturer and industry standard best practices.

Switching technology in the MDF and IDF will support the following;

I. Full range of VLANs

II. Port management

III. Access restriction

IV. SSH remote administration access

V. Standards based routing will be supported

VI. Extended warranties should be taken out to secure the devices

VII. Same day replacement support, hot spares of key components should be considered

VIII. Devices should support remote firmware management from server based management platform

IX. Vendor specific configuration management tools will be applied to backup device configuration

X. System wide VLAN management (e.g. VTP)

XI. Mix of powered and non-powered ports depending on requirement

XII. MDF core devices to support 40Gb Ethernet throughput

XIII. Ports not in use will remain in a state of shutdown until required

15.9.10 Servers, User Devices - Hardware Standards

General

I. System software vendors often specify their hardware requirements for "optimal" performance. However, the IT departments of facility companies offer alternative opinions about optimal performance and must be consulted prior to any acquisition.

The reasons for this are;

(a) Vendor specifications for server hardware are often lower end to appear less costly

(b) Vendors often insist on dedicated servers when their software could sit in a server array

(c) Vendors sometime do not like their software to be in a virtualised server environment. Facility companies have a different view because their objective is to provide excellent system performance and hardware efficiency at the best price possible

(d) Racks, server arrays, workstations, printers and peripheral devices deployed in the Data Centre will be configured by the Owner's IT department in co-operation with vendors

Where required, it is essential that the dimensions and models of all hardware (along with locations) are provided to M&E engineers, architects and interior designers to accommodate them aesthetically within design and to ensure they can be operated efficiently.

15.10.1 Telecommunication Standards

Mixed-Use building requirements is a very challenging environment for managing telecommunications. For SIP (session initiated protocol) iP PBX systems, software and the related infrastructure, it is imperative that facilities use telecommunications vendors who understand and operate in the environment. It is not only important that the systems are configured to maximise functionality and uptime, but also to deliver acceptable ongoing operating expenses. Reputed manufacturers shall have facility specific software solutions providing the functionality required.

I. Communications play a key role in the hospitality industry and it is very important that the property has the utmost confidence in its vendor to ensure that the system remains operative 24 hours a day. Often this can be the difference between life and death in emergency situations

II. When evaluating the services of a telecommunications vendor, there are a number of questions requiring satisfactory answers;

(a) Which other clients in the region does the vendor support?

(b) Obtain references from other facilities of similar size and configuration

(c) Where is the vendor’s closest support office located?

(d) Do they have two-to-four-hour response time 24 x 7 x 365

(e) Who will be servicing the account?

(f) Vendors will sometimes subcontract the support of if they do not have their own support office in the area. If this is the case, it will be important to obtain additional information on the subcontractor.

(g) How well equipped is the local support office?

(h) If a major component of the system were to fail, how long would it take to replace?

(i) In most cases, it is advantageous for the local support office to carry a complete set of replacement parts for the system to reduce downtime

(j) Are the support technicians certified to support your system?

(k) Are they expert with the configuration, functionality and integration of all the key components of the system?

- PBX
- Voice mail
- Call accounting
- ACD

15.10.2 Telephone (iP PBX) System

I. An iP PBX provides voice and data over an internet protocol (IP) internal structured data network. The clear benefits of this technology are;

(a) IP PBX are easier to install and configure than a proprietary system

(b) Easier to manage through web based user interfaces

(c) There are significant cost savings if VoIP service providers are available to purchase SIP trunks

(d) Eliminates dedicated phone wiring by using the Structured

Cabling network

(e) Eliminates vendor "lock-in".

(f) IP PBX's are based on open SIP standards, so most any SIP hardware or software can be used without loss of functionality which is particularly relevant when choosing handsets

(g) Scalable without substantial proprietary add on costs

(h) Phone functions can be integrated with business applications at the desktop

(i) Hot desking and roaming features readily available

(j) Significant savings in management, maintenance and carrier costs

II. Configuring the requirements for an IP PBX is determined by the nature and size of the facility and the following must be taken into consideration;

(a) Number of "extensions" – public areas, restaurants, leisure facilities, corridors, back of house service areas, administration etc

(b) Number of "trunk lines" be they plain old copper or SIP trunks

(c) Number of reserved trunks for "private line" demands

(d) Minimally eight (8) "power fail" analogue trunks for emergency fail save procedures should there be power outages or major system failures

(e) Number of telephone operator attendant stations be they a fixed position or floating determined by time of day

(f) Individual to each user – headsets for telephone operators

(g) Functional low cost back of house corridor phones to fully functioning admin phone**s**

III. Many of the facility's employees need to communicate internally and between themselves to provide operational efficiency and immediacy of service required in a facility. The suppliers/vendors who are invited to tender must be requested to propose such a system integrated with the iP PBX for a minimum of 50 users. This solution may be one or a combination of;

(a) **Paging System**

(b) DECT (Digital Enhanced Cordless Telecommunications)

(c) Localised cellular network

(d) Wireless Voice over IP

Whichever system is selected it MUST interface with the Fire Alarm, Building Management Systems and Intruder Alarm Systems so that a response team member can be notified of any issue reported by one of those systems

I. The IP PBX's Facility Specific Software must have the following features;

(a) Caller name display, language designation received from the Property Management System interface

(b) Recorded announcements (individual & broadcast) e.g emergency calls in a minimum of 6 languages

(c) Music on hold

(d) Automatic Call Distribution with management reporting.

(e) Classes of Service and ring down to operator

(f) Speed Calling to a minimum of 100 numbers

(g) Least cost routing of trunk calls where obtainable (availability determined by in country PTT)

(h) Call waiting to not conflict with voice messaging service

(i) Operator and station access to instigate internal paging/communication with employees

(j) Trunk Traffic statistics recording and reporting

(k) Power Failure Transfer – 8 analogue 2 way trunks.

(l) End location determined by management

(m) Fully Redundant CPU with failure automatic switching without loss of service or data

(n) Property Management Interface

15.10.3 Voice Mail (VM)

The proposed system is to include an integrated facility voice messaging system to manage administrative voice messages.

This section is to provide a minimum requirement specification to assist in the product evaluation of Voice Messaging systems and to ensure that a consistent level of brand quality and service is provided throughout the facility. It is critical to customer satisfaction that the installation of a Voice Messaging system has appropriate features and functionality.

I. The Property Management System (PMS) integration to the Voice Messaging provides the necessary level of integrity to the system.

III. When a user accesses the VM system from outside of their offices, they will be required to input Two-Factor Authentication in order to retrieve their messages.

IV. Message waiting lights will be automatically activated upon voice mailbox message or PMS notification.

V. The VM system must be tightly integrated with the Facility's Telephone System.

VI. The proposed VM must allow for direct DID/DDI access to user mailboxes.

VII. The VM system must accommodate at least 4 language sets. The VM vendor must present a list of Facility references that have installed their system to date and references must be checked before any purchase of an VM system.

15.10.4 Call Accounting

I. Specialist telephone call accounting systems are a vital part of the telephone services and must be included as an integral component of communications control. The system must be fully interfaced with the iP PBX and the facility's PMS and must include an off-line alarm.

II. The following are key functions identifying why call accounting is a specialist requirement:

(a) Full two-way PMS Interface for automatic posting of user telephone call charges

(b) Billable user call records immediately sent to Property Management system

(c) Multiple billing methods and rates, billing algorithms

(d) Exception charging

(e) Dynamic DID number allocation from the PMS to iP PBX

(f) Unique billing for Concessions, conferences, meetings

(g) The user will have available an itemisation of all calls placed and the charges where appropriate

(h) Direct dial billing

(i) Trunk signaling (i.e. answer supervision)

(j) No single failure can cause complete loss of data. (Including disk drive head crash)

(k) Automatic system restart from a power failure.

(l) Statistical analysis of outbound traffic by trunk or summary

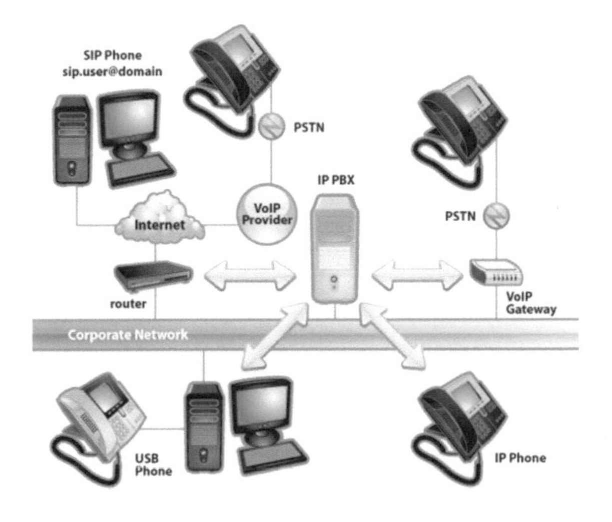

15.10.5 End User Devices

I. Telephone - Public area courtesy telephones must have the following features;

(a) Single line corded IP

(b) Dialing key pad on handset

(c) LCD display

(d) Message waiting light

(e) Redial key

(f) Battery backup for power outages

(g) Desk or wall mountable

(h) Optional colour schemes to coordinate with interior design

II. Administration Station IP telephones can be supplied by the PBX manufacturer but it is preferable that ACD Headsets, Operator Headsets and Meeting Room Speaker phones are a specialist selection.

15.11.1 WI-FI

Smartphones and tablets are changing the way society communicates with each other and accesses content. Forrester Research labels this as "the mobile mind shift— the expectation that a person can get what they want in their immediate context and moments of need." Simply put, people turn to their mobile devices for answers to their questions. A strong wireless infrastructure foundation is needed to accommodate these mobile moments. But all of that mobility relies on a strong infrastructure mix of wireless equipment, fiber, and cable.

With the explosion of Wi-Fi enabled devices and rising user expectations, provisioning strong, fast, reliable Wi- Fi should be given serious attention. Wi-Fi deployment requires a considerable level of planning to ensure that there are no dead zones, disconnects, jitter issues and slow response/ refresh.

The Wi-Fi network footprint is expected to cover the entire Multi-Use facility. The footprint within a space is expected to cover the entire area to a minimum signal strength of -65db. Wi-Fi standards should at least meet the most recent 802.11 ratified standard. At the time of print this is 802.11ac. Backward compatibility for 802.11n devices and earlier shall also be addressed.

15.11.2 Design Overview – WIFI and Fixed Wire

General

The provisioning of Wireless High-Speed Internet (Wi-Fi) services in the multi-use building facilities will require a in-building solution such as a Distributed Antenna System (DAS), building amplifier, or other small-cellular solution to boost coverage and capacity. Wireless carriers own the RF spectrum required to bring cellular connectivity inside a building so the designers must work with carriers and their partners (often called third-party neutral host providers or systems integrators) to achieve coverage.

The designer must create an accurate electronic Wi- Fi design, knowing what applications the WLAN will support, who will have access to the network and where they are likely to roam throughout the building. The latest Wi-Fi surveying and planning tools must be used to electronically pre-design the WLAN.

To avoid compromised 11n performance in a mixed-mode network, use an AP with two radios: one tuned to the 5 GHz band and the other tuned to the 2.4 GHz band. Use the 5 GHz band for 802.11n WAPs and clients, creating a "pure 11n" network in that band.

Allow legacy 802.11g clients to communicate with a 2.4 GHz "G" radio in the other slot.

The installation may consist of fiber or UTP cable, Category 6, from Main Computer Room or Intermediate Data Cabinets to locations throughout the facility for the mounting, installation and servicing of Wireless Access Points (WAPs). All reference to conduit, cable placement and termination requirements for Category 6 cabling in this document must be adhered to for installation of Wi- Fi cabling. The scope of this definition relates only to the design and installation of the Cabling system for this Wi- Fi network. Reference to WAPs in this document is only intended to give a point of reference for design of the Cabling system.

WAPs should be concealed from public view, or mounted in a subdued and aesthetic manor as well as secured from tampering if in an accessible area. WAPs will be permanently accessible after installation, to facilitate repair or replacement. WAPs will not require access to electrical power near their installation, but may utilize it if available. Terminated cables for the Wi-Fi network must be emplaced throughout the Areas of Coverage in sufficient density and proximity to each other.

A 4.5m to 6.0m piece of cable called a service loop must be left in the ceiling (Figure 48.3) in case a WAP later needs to be moved slightly to tune coverage or avoid interference from other RF devices, such as wireless phones and microwave ovens.

The characteristics of the facility should be factored and would include;

Location, category, size of the facility and its retail spaces, office spaces, residences, the public areas, exhibition & meeting spaces, leisure facilities, land to perimeter, back of house, corridors, stores, car parks etc

Construction materials, geographical location, interferences

Specialist expertise is required to design, install, test, monitor and support a Wi-Fi/ fixed wire installation

Identify all the equipment (such as switches, routers, firewalls) and tools (e.g authentication, encryption and intrusion) necessary to identify, trap and nullify intrusion/ security threats

15.11.3 General Standards

I. A dedicated data connection, with redundancy failover, is required to the Internet Service providers' equipment with a minimum bandwidth of 100 mbps. Data provisioning will be provided over a fibre based leased line on a business class service and backed by a service level agreement (SLA) to 100% uptime. Any loss of service should earn service credits as compensation.

II. The service provider is expected to provide proactive remote monitoring to ensure the performance of the network, identify abnormal network activities and address system performance issues such as:

- Traffic management
- Channel clash

III. Where internet service is provided over the buildings structured cabling, clear demarcation of responsibility and roles must be identified so as not to hinder support of the system.

IV. The internet service may run on a separate cabling infrastructure to that providing data, voice and building services

V. Wireless network throughput shall bet least 1mbpsConference/hotspot access for event delegates etc.

VI. Access must be instant, with no additional software app. required by the user

VII. Wi-Fi Network security must be provided by the HSIA provider

VIII. The service will not hinder users from access to any legitimate sites, email or VPN services. Access will only be denied in adherence with local laws and where the facility as the ISP can be held liable for the users' actions

15.11.4 Wireless Access Pointsk

I. The wireless installation must conform to the latest IEEE 802.11ac standards addressing multiuser MIMO communications and potentially 160 MHz wide channels. The positioning and number of Access Points should be determined by a site survey producing a heat map of signal strength and coverage.

II. Account for equipment that causes radio interference. This will include building construction materials and wireless penetration from neighboring building

III. The location of Access Points should also take into consideration the heat output of the WAP and ventilation required of the enclosure/location

IV. Access points will be managed by Power over Ethernet (PoE) enabled network switches/controllers

V. Access Points are to be positioned discretely within the space or ceiling mounted in corridors/public areas

VI. Access Point indicator LED's must have the ability to be turned off in order to make them more discrete, especially within Residential Apartments.

VII. Utilise Access Points that enable higher end mobile devices in the 5 GHz spectrum to move out of the still crowded 2.4 GHz spectrum. This greatly assists in balancing load

VIII. Access point capacities, coverage and traffic requirements need to be identified

IX. Peak usage periods should be identified and allowed for

Post implementation, a thorough survey should be conducted to ensure optimal locations for access points and that any weak or dead zones are rectified

15.11.5 HSIA Controllers

I. HSIA controller equipment will be located in the secure Mechanical Electrical Room/Data Centre from where all wireless access points will be centrally managed.

II. Features will include;

(a) The controllers will have redundant components to ensure continuous service. For example, power supply units, network interface cards

(b) Properties of the controller will include but are not be limited to;

- Adaptive wireless meshing
- Sophisticated user access controls
- Automatic traffic redirection
- Integrated Wi-Fi client performance tools
- Simple but comprehensive guest networking functions
- Rogue WAP detection and advanced Wi-Fi security features
- Flexible WLAN groups
- Extensive authentication support
- Robust network management

III. The controllers will have adequate capacity to manage the required number of access points identified as necessary to provide the superior level of service required.

IV. The controller will support captive portal features

15.11.6 Service Management Platform

I. The high-speed internet service provider must supply an intelligent management platform that is specific to the Mixed-Use Building industry. The platform should be capable of sensing channel clash and managing user authentication via passwords, ticketing etc. The system must manage security to block denial of service and spoofing attacks.

II. There should be Wi-Fi protected access based on Advanced Encryption Standards along with strong message and integrity checking.

III. The system should be capable of creating private networks within each space, allowing a user to connect between multiple wireless devices while staying separate and secure from other users on the network.

IV. The HSIA system must be configured to present the user with a Terms of Service page that the user must accept prior to being permitted to use the HSIA service. The HSIA provider is responsible for providing the Terms of Service that will contain the network security risks associated with the use of HSIA.

V. Once the user has accepted the Terms of Service, a re-directional link to a Start Page/landing page is required. This page must be displayed as the gateway to the Internet and include relevant facility information

15.11.7 Bandwidth Management

The Service Management Platform should be capable of;

I. Identifying individual high bandwidth users and be able to manage that bandwidth by restriction or advising that additional charges apply for such utilization

II. Issue user access policies to manage users on individual Service Set Identifiers (SSID) to ensure fair use

III. Dynamic and configurable client load balancing to optimize overall WLAN system performance

15.11.8 Security

The Service Management Platform and network should make use of available best practices and technologies to protect the network and its users. It should be capable of;

I. Rogue Access Point detection – The system should be able to detect unauthorised WAP's “installed” on the network

II. Client blacklisting – Clients who are detected as sending spam or acting maliciously should be blocked from the network until further investigation deems it safe allow access

III. Denial of Service (DoS) attack prevention – Tools should be employed to detect and prevent DoS attacks

IV. Password guessing protection should be employed to protect the management servers and WI-FI controllers

V. Connected clients shall be isolated from other clients to protect against peer-to-peer snooping

VI. L2-4 ACL (Access Control Lookup) support

VII. Given the risk associated with network security, it is recommended that a facility ensures that all guests are made aware that there is a risk associated with the use of HSIA and that the user assumes that risk

15.12.1 Internet Support

Professionally supporting the users' needs and enquiries is essential. Device problems sometimes occur in accessing and using the facility's internet service. The facility operator may operationally choose to field the first support call to an “cyber assistant”. However, the user or the cyber assistant must be able to access a professional support service with standards as follows;

I. 24/7/365 technical support through a reputable third- party company that provides HSIA support as a part of its core business.

II. A “live” person must be available to address all technical support calls

III. An initial automated tele-prompt service is acceptable only if a live person is reachable via touch prompt. Message and call-back service is not acceptable.

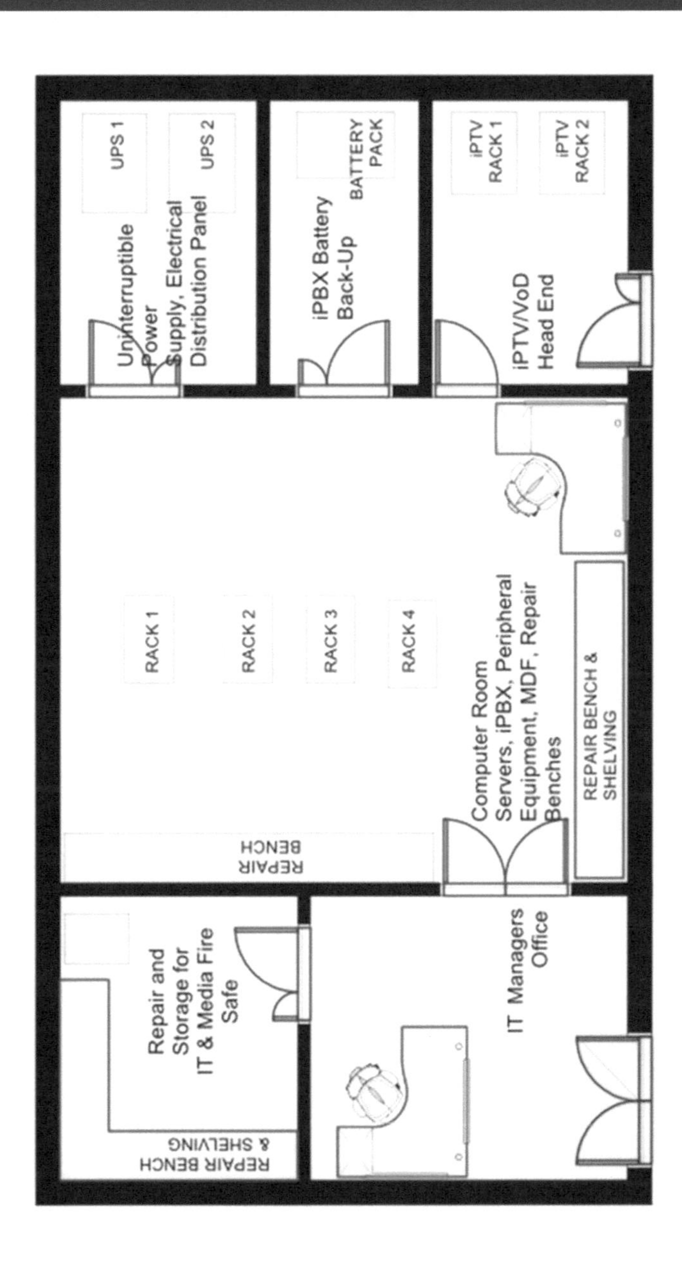

15.13.1 Building Internet of Things

General

Technology is becoming increasingly intertwined with our day-to-day routine. With advancements in technology, not only have we become smarter, but so have the devices we rely on. Smarter devices enable a new type of infrastructure, an intelligent network built on the Internet of Things (IoT).

The design development team need to anticipate IoT trends whilst designing Multi-Use Building facilities. In a fully IoT-enabled building, the convergence of power, light and data over a common cabling system for a variety of applications should be achieved.

It is essential that the building facilities being designed will be able to support IoT devices without needing to be redesigned. When considering a facility design that supports IoT, the designer must anticipate the technology trends for key building components with different lifespans.

The availability of real-time data that mechanical/ erlectrical systems must be utilized to maximize reliability, equipment life, enhance safety and comfort, and lower operating costs.

Internet vs. intranet, wired vs. wireless

Besides accessing data from across the internet, IoT devices may access data from a local network (intranet) or exchange data locally with a peer device.

Depending on where the device is and where the data is located, IoT devices may be wired or wireless.

Power over Ethernet protocol or some other Class 2 or Class 3 power source may be used in the network room, for wired IoT devices in need of a power source. The building Wi-Fi network should facilitate the communication between wireless devices.The designer should investigate the possibility of using a higher density of wireless access points with a higher bandwidth and needing more power.

Since all the cables go back to the network communication room, the cabling takes space in the network communication rooms and in the pathways getting back to the rooms. Both the pathways and the rooms should be designed with growth in mind. A good rule is to allow for 100% growth for the horizontal cabling spaces (cables going from a device back to the network communication room) and 50% growth for the backbone cabling spaces that interconnect the network communication rooms. This will allow room for growth in both the network communication rooms and cable trays.

The designer should design a room for the equipment being installed on day 1 as well as for equipment to be added over the lifespan of the building. The network communication rooms should house all network equipment, whether it is for private use or public-access networks.

If a building owner/operator decide not to allow public IoT devices to access their private network or even their guest wireless private network (mobile IoT devices can communicate with multiple types of wireless protocol), they should be able to communicate over a cellular network. if a public-access Wi-Fi network is not available, the public cellular networks may be extended into a building through a distributed antenna system. This equipment takes space, requires power, and should be integrated into the network communication rooms.

The convergence of technology disciplines (telecom, security, building automation, lighting etc.) has an impact on the architectural layout of the building and the concentration of loads on the electrical and mechanical systems. Additionally, artificial intelligence and automated response devices should be interconnected and more thoroughly integrated into the surroundings or building systems.

Mesh networks shall be employed to interconnect IoT devices, where every node interconnects with a nearby node to pass data. Mesh networks should be set up to be self-recognizing and self-healing. Where the Wi-Fi network is unable to reach spaces in the building, wireless interconnection of nearby nodes, IoT devices can act as digital repeaters.

For example, the ZigBee protocol could be a wireless data solution characterized by secure, reliable wireless network architectures.

The Benefits of the Building Internet of Thing

Unified communications solutions help eliminate cost and complexity and accelerate the pace of innovation. The benefits of a state-of-the-art business communications and collaboration platform without any of the day-to- day operational hassles of a traditional premises-based product is essential in today’s Multi-Use Buildings.

Improved life/safety

- Systems are continuously monitored
- Potential for improved emergency evacuation procedures

Increased comfort

- Systems working optimally, and problems are caught before they can be noticed
- Personalization of the workplace
- Increased control of environment, and an ability to tune the environment to the individual occupant's preferences
- Improved wayfinding
- Internet of Things devices and apps can support wayfinding from the parking garage right to the destination point in a mixed-use building.
- Improved occupant engagement with facility initiatives, such as recycling or load shedding.

Robust communications networks are considered a pre-⅓ requisite by prospective tenants

Building Internet of Things provides functionality to start-⅓ups and smaller companies which was previously mainly available to larger enterprises, such as paging systems and receptionist/office manager functions

Connected Device: These refer to a wide variety of "Things" such as smart thermostats, telephones, security cameras, security badges, and more. These devices should support various applications by collecting useful data with the help of various existing technologies and then autonomously communicating the data between other devices to make enhancements to the surrounding environment without human interaction. The connected devices shall leverage IP and many be ready for power over ethernet (PoE).

Enabling Infrastructure: This refers to the structure beyond the "Things" that make the IoT philosophy possible. The infrastructure should include the platforms which facilitate a common language for all devices to communicate freely, the "collect and act" scenarios that are the essence of the IoT movement and the "enablers," such as PoE delivery, WAPs, gateways and edge devices. The cabling infrastructure that makes up the "physical" deployment of the IoT installation shall be at the core of these systems. A structured cabling infrastructure should provide the required foundation to support these applications.

Applications that Leverage Internet Protocols:

- Wireless Access Points
- Security Cameras
- IP Phones
- Intelligent LED Lighting
- Occupancy Sensors
- Climate Sensors
- Access/Building Automation Controls
- Digital Signage
- Sound Masking
- Visible Light Communication (i.e. Li-Fi)

Cabling and Connectivity Considerations:

Cabling must support adequate power throughput and efficiency in addition to the heat dissipation capabilities.

Connectivity must be robust, durable and provide power headroom for current carrying capacity. Arcing shall be prevented.

A hybrid approach of both a centralized and decentralized deployment may be necessary to support the variety of applications and industry requirements.

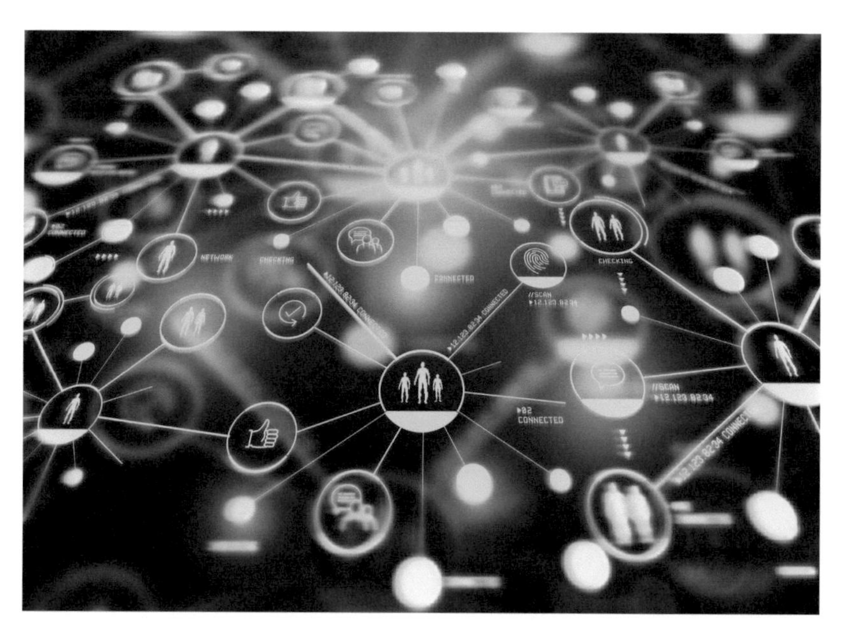

SECTION 16.0

Testing & Commissioning

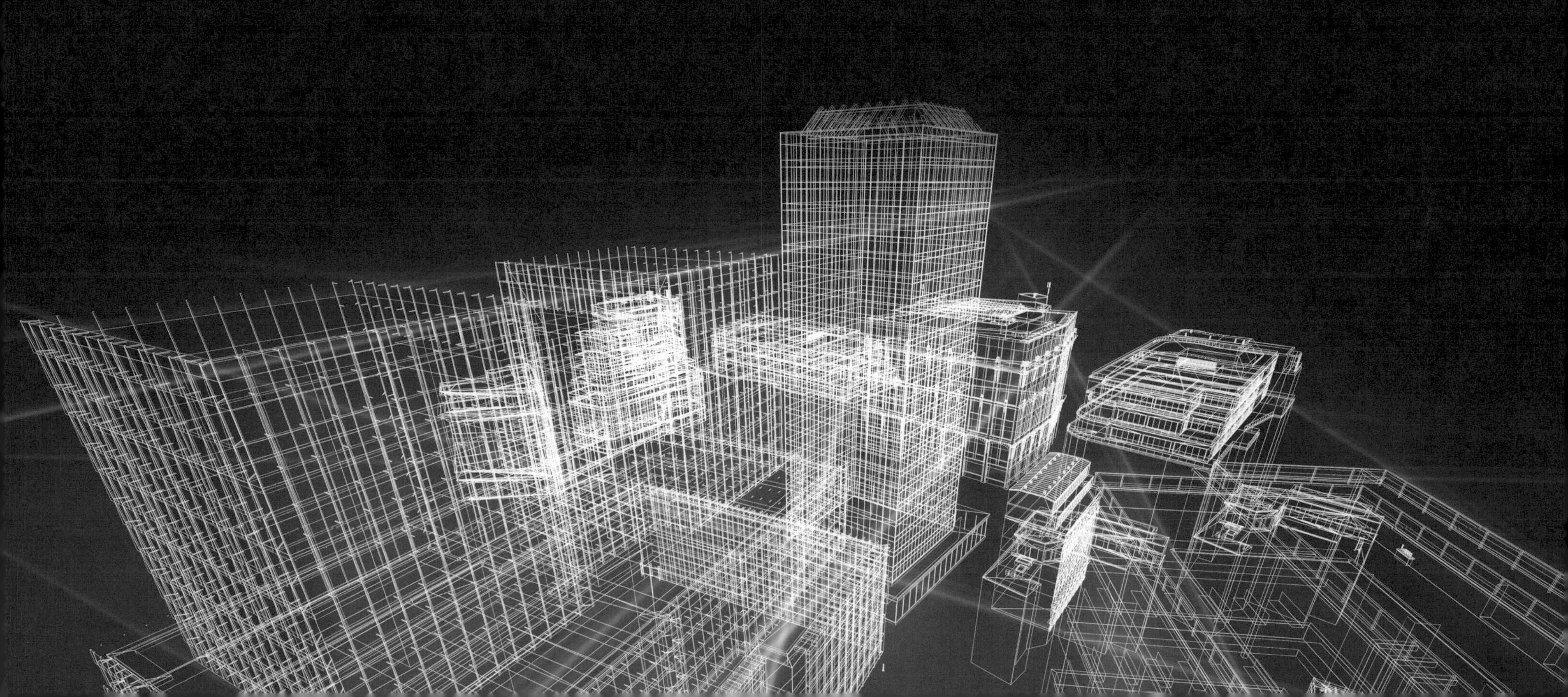

16.1.1 Factory Testing (Equipment)

Witness testing should be carried out at factory, as required by the appropriate National and International standards. If the performance of the equipment offered is not satisfactorily demonstrated, then further tests should be carried out until the Owner is fully satisfied that the equipment offered meets the requirements. These further tests should be at no extra cost to the contract.

16.1.2 Site Testing

A visual inspection should be made of the installed equipment before any tests are carried out to confirm the following:

- Correct selection and erection of materials and workmanship
- Signs of damage so as to impair safety

During construction of the works, the installer should undertake all necessary tests to ensure compliance with the regulations and specifications relating to the works.

Upon completion of the works, the whole installation should be subject to the tests required by the regulations and specifications relating to the works. These tests should be witnessed by the IT Department to their full satisfaction.

16.1.3 Commissioning

Professional commissioning of the systems is essential to ensure:

- The system's performance in accordance with the design
- User comfort is not jeopardized
- Operating and maintenance costs are minimized

It is therefore critical to the ongoing operation of the facility that this key task is carried out correctly in a thorough and professional manner. The Owner reserves the right to bring in an Independent Specialist Commissioning Company if this task is not being carried out directly.

The installer and designer should verify the correct operation of the complete systems, including items of plant, to ensure that they are operating in accordance with specified conditions and that the systems achieve the environmental conditions required. This should include proving of equipment installed under full load conditions. Where electrical loads are not inherent in the installation, temporary electrical loads should be provided.

Upon completion of all testing and commissioning, the installer and designer should provide two signed copies of the commissioning certificates and submit to Owner within 14 days of the results being obtained. Signed copies of the certificates should be installed within the operating and maintenance manuals.

The installer should provide,at his own cost,all water,power, fuel,labor,etc. necessary for all testing and commissioning.

16.1.4 Method Statements

Prior to all testing, the installer and designer should issue detailed Method Statements prior to carrying out the works. These method Statements should contain the following information:

- Health and Safety Issues
- Instruments to be used and their calibration certificates
- Sequence of tests to be carried out
- Documentation which will be provided to record results

Full and detailed Method Statements should be provided for all pre-commissioning, setting to work, commissioning, testing and handover procedures. These Method Statements should include the following information:

- Logic diagram of the process
- Outline program (detail to be added as the document develops)
- Copies of all checklists, record sheets, etc.
- Permit to work systems and documentation
- Details of work by others affecting progress of work detailed
- Proposed off-site testing
- Proposed completion sequence
- Proposals for quality control
- Handover, demonstration and training

The Owner reserve the right to request additional Method Statements for any aspect of the works, at no additional cost to the contract

16.1.5 Program

A program should be submitted to detail all key aspects of the commissioning, including but not limited to:

- Submission of Method Statements/procedures
- Utility connection dates
- Individual system commissioning dates
- Witnessing dates for authorities/Owner

The following simplified program provides details of the expected timescales, but this should be prepared in detailed form, specific for each project.

Sample: Commissioning Program – MEPF Ser vices								
Task	Week Number							
Appraisal of Method Statements/ Procedures								
Utility Services Connected								
Mechanical Systems								
Check / Clean / Flush								
Electrical Checks / Run								
Water Balance								
Air Balance								
Chillers / Pumps								
Boilers / Pumps								
Controls / BMS								

Sample: Commissioning Program – MEPF Ser vices								
Task	Week Number							
Electrical Systems								
Transformers / Main MV Panel								
LV Panels								
Power								
Lighting and Controls								
PA / CCTV / Security / Commission								
Fire System								
Fire Alarm and Detection								
Smoke Control								
Fire Hydrants and Sprinklers								
Misellaneous Fire Extinguishing Systems								

Sample: Commissioning Program – MEPF Ser vices								
Task	Week Number							
Plumbing and Drainage								
Flush / Disinfect HWS								
Flush / Disinfect CWS								
Water Treatment								
Booster Pumps								
H & CWS Balance								
Grease Traps / Clean Drains								
Elevators (60 days) (Constr uction Use)								
Owner/ Authority Witnessing								

16.1.6 Record Documentation

The installer should prepare full operating and maintenance manuals and As Fitted drawings in both English and local language, if applicable. Draft copies of these documents should be issued to the Owner for comment a minimum of six weeks prior to practical completion.

All record drawings should be provided in electronic form (CD ROM, Windows), in addition to paper copies. Drawings should be provided on the CD ROM to the latest Auto CAD Release version and should be prepared using agreed CAD layering convention.

Three (3) paper copies and three (3) CD ROM copies should be provided, with three (3) additional copies of each in the local language, if applicable.

16.1.6.1 Record Drawings

Record drawings should be produced, which should include the following minimum information:

- The location of all public service connections, within the contract, whether installed by the contractor or by the appropriate Authority, together with points of origin and termination, size and materials of pipes, line pressure, flow and other relevant information.
- The layout, location and extent of all sub-mains cables and piped services together with all isolation points, valves, test points, etc.
- Location, identity, size and details of all controls equipment
- The layout, location and extent of all air ducts, including dampers, silencers, and air flow quantity, etc.
- The location and identity of each room, including spaces housing plant machinery air apparatus
- Detailed general arrangements of all machinery spaces, air handling plant rooms, tank rooms, electrical switch rooms, etc. including location, identity, manufacture, size and rating of all equipment.
- All necessary sections, elevations, isometrics and schematic of all plant spaces.
- All controls and wiring diagrams should be provided.
- Floor layouts should be provided at a scale of no less than 1:50. Plant areas and equipment rooms should be at a scale of not less than 1:20.
- Each record drawing should indicate:
    - Name of the contract and the area of the building
    - Description of the drawing, unique number and scale
    - Name and address of the installer
- All record drawings should be signed and checked by the installer and designer in accordance with agreed quality control procedures.
- In addition to Record Drawings, the following wall-mounted, glass covered schedules and schematic layouts should be provided. These should be located in plant rooms and any other appropriate locations:
- Schematic drawings of all systems and circuit layouts showing identification and duties of equipment, numbers and locations, controls and circuits.
- Control schematics

- All items required under statutory / other regulations
- Emergency operating procedures and telephone numbers for emergency call out services

All the above is to be submitted for approval prior to erection.

16.1.6.2 Operating and Maintenance Manuals

The operating and maintenance mmanuals should incorporate the following minimum information:

- A section containing an introduction, abbreviations, health and safety and working notices, etc.
- A section containing full description of each system, together with main plant components, locations, mode of operation of automatic control systems, etc.
- A section containing plant technical data for all items of equipment
- A section describing in detail, operating procedures necessary for start-up, running and shut down of equipment and any system
- A section describing maintenance operations on a daily, weekly and monthly basis for each item of equipment and any system
- A section describing the emergency procedures to be adopted by personnel engaged on the operation and maintenance of the systems with respect to fire, first aid, failure, etc.
- A section describing recommended action on plant malfunction
- A section listing recommended spares and lubricants
- A section containing all the record drawings
- A section containing all test certificates and commissioning reports
- A section comprising a list of manufacturers, including addresses, telephone numbers and equipment supplied

A. Design Compliance Checklist

This section includes the following Design Compliance Checklists which should be completed by the Designers and submitted to Owner as indicated.

Summary schedule of Information and Stage to be submitted.

(a) **Checklist 1 - Design Stage Compliance Verification**
(b) **Checklist 2 - Site Location / Summary Details**
(c) **Checklist 3 - Climatological Data**
(d) **Checklist 4 – Electricity**
(e) **Checklist 5 - Gas Ser vice**
(f) **Checklist 6 - Potable Water**
(g) **Checklist 7 - Sewage Disposal**
(h) **Checklist 8 - Fire Protection**
(i) **Checklist 9 - Boiler Fuel Options**

SUMMARY SCHEDULE OF INFORMATION AND STAGE TO BE SUBMITTED				
SCHEDULE OF INFORMATION				
Required Information	Design Period		Pre-Contract Works	Handover
	Preliminary Drawings	Final Scheme	Detailed Construction	As Built Record Drawings/Manuals
Mechanical and Electrical Services Schematic & Load Schedules	Outline	Detail	Detail	Detail
Mechanical and Electrical Services Drawings & Specifications & Site Plan	Outline	Detail	Detail	
Energy Conservation	Calculation/ Appraisal	Detail	_	_
Drainage Plans and Details	Preliminary	Outline	Detail	Detail
Fire Safety Systems and Specifications	Outline	Outline	Detail	Detail
Fire Safety Strategy Proposal	Preliminary	Detail	Final	
Calculations	Outline	Detail	_	_
Technical Submission for Plant & Equipment	Preliminary	Detail	Final	_
Schedule of Deviations	Outline	Detail	_	_
Checklist 1 - design Stage Compliance Verification			_	_
Checklist 2 -Site Location / Summary Details			–	–

SUMMARY SCHEDULE OF INFORMATION AND STAGE TO BE SUBMITTED				
SCHEDULE OF INFORMATION				
Required Information	Design Period		Pre-Contract Works	Handover
	Preliminary Drawings	Final Scheme	Detailed Construction	As Built Record Drawings/Manuals
Checklist 3 - Climatological Data			–	–
Checklist 4 - Electricity			–	–
Checklist 5 - Gas Service			–	–
Checklist 6 - Potable Water			–	–
Checklist 7 - Sewage Disposal			–	–
Checklist 8 - Fire Protection			–	–
Checklist 9 - Boiler Fuel sOption			–	–
Checklist 10 - Sample Mechanical Load Schedules			–	–
Checklist 11 - Sample Electrical Load Schedules			–	–
Checklist 12 - Building Air Balance			–	–
Note:				
Outline: All information needs to be indicated but not fully described				
Detail: The complete information to fully describe the works				

CHECKLIST 1: DESIGN STAGE COMPLIANCE VERIFICATION				
	Item	Compliant	Non-Compliant	Comments - If Non-complaint
1.0	Aims & Objectives			
2.0	Fire Safety			
	~ Detection and alarm			
	~ Suppression			
	~ Emergency Lighting			
	~ Gas Detection			
	~ Cause and Effect			
3.0	Mechanical / HVAC			
	~ Design Criteria			
	~ System Selection			
	~ Energy Efficiency			
	~ Load Schedule Completed			
4.0	Electrical			
	~ Power Supply			
	~ Emergency Power			

CHECKLIST 1: DESIGN STAGE COMPLIANCE VERIFICATION				
	Item	Compliant	Non-Compliant	Comments - If Non-complaint
	~ Lighting			
	~ LV System			
	~ Load Schedule Completed			
5.0	Plumbing / Sanitary			
	~ Water Quality/Treatment System			
	~ Cold Water Reliability			
	~ Drainage			
6.0	BMS			
	~ System & Points Schedule			

CHECKLIST 2: SITE LOCATION & SUMMARY DETAILS	
ITEM	
Country:	
Site Address:	
Location (City/ Combined Development)	
Latitude/Longitude:	
Proposed Facilities	
Total Area – Occupied	
Spaces/Storage:	
Total Area – Plant	
Total Area – Car Parking:	

CHECKLIST 3: CLIMATOLOGICAL DATA			
1.0	Annual Rainfall:		
	Maximum Rate:		
2.0	Number of annual cooling degree days:		
3.0	Summer Design Temperatures (1%, 2½% – DB/WB):		
4.0	Average Wind Speed and Direction for each season:		
		West	
		South	
		North	
		East	
5.0	Number of Annual Sunshine Hours;:		
Remarks:			

CHECKLIST 4: ELECTRICITY			
1.0	Obtain full network characteristics, including transformer type and site, fault current, rate schedule, rules and regulations, national and local codes, etc., and attach		
2.0	Voltage Stability		
3.0	Frequency Stability:		
4.0	Number of Brownout Hours:		
5.0	Number of Blackout Hours:		
6.0	Identify Major Causes of Ser vice Disruption:		
7.0	Are two independent ser vices possible?	Yes	No
8.0	Adequacy of system over next 5 years:		
9.0	Application lead time:		
10.0	Available tariffs:		
11.0	Separate light and power tariffs or flat rate:		
12.0	Unit costs, night/day charges, season variations:		
13.0	Demand charges and when applied:		
14.0	Power factor requirements:		
15.0	Controlling agency		
16.0	Is frequency change contemplated?	Yes	No
Remarks:			

CHECKLIST 5: LP GAS SERVICE			
1.0	Is utility, piped-in LP Gas available:	Yes	No
2.0	Alternative Gas Sources		
3.0	If LPG, is bulk delivery available	Yes	No
4.0	Cylinder size:		
5.0	Calorific Value (CV), if known		
6.0	Cost		
7.0	Is supply adequate?	Yes	No
8.0	Are LPG installations subject to local codes/ practice?	Yes	No
9.0	Is copy of code/regulations attached?	Yes	No
10.0	Is LPG delivered by tank truck:	Yes	No
11.0	Indicate size of cylinders:		
12.0	Indicate cost of LPG:		
13.0	Name applicable codes:		
Remarks:			

CHECKLIST 6: POTABLE WATER SERVICE			
1.0	Primary Supply by:		
2.0	Alternate Sources:		
3.0	Distance to Mains:		
	Size		
	Pressure		
	Elevation		
4.0	Age and Condition of Mains if known		
5.0	Largest tap permitted:		
6.0	Are Dual Services available	Yes	No
7.0	Limits on peak demand	Yes	No
	Limits on peak consumption	Yes	No
8.0	Rate(s) applicable:		
9.0	Overall water quality		
	Laboratory Report attached:	Yes	No
	If no, when is this expected:		
	Turbidity:		
	Odour and taste:		
	Total hardness for each season:		
	pH for each season:		
	Bacterial count:		
	Chlorine residual:		
	Corrosive or scaling tendency of water. Reference is made to Langelier & Ryznar Indices.		
10.0	Reliability: total hours shut down annually:		
	Identify major cause(s) of ser vice disr uption:		
11.0	Application Lead Time Agency:		
12.0	Adequacy of System over next 5 years:		
13.0	Summarise treatment proposed to achieve Owmer established water quality requirement as detailed on the next page.		

Water Purity			
Potable water quality: Permissible concentrations			
Physical-chemical limits and maximum permissible concentrations in parts per million (ppm).			
Parameter	WHO Standard	Desirable Levels	EU Standards
Temperature (°C)		10 - 15	<25
pH.	6.5 - 9.2	7 - 8	6.5 - 8.5
Conductivity (mS/m^3)		400	400
Chlorides	250	<50	250
Sulphates	200	<250	250
Hardness (as CaCO3)		<100	100
Magnesium		<50	50
Arsenic	0.01	0	0.04
Cyanides	0.01	0	0.05
Aluminium		<0.2	0.2
Total Dissolved Solids (TDS)	1000	<500	1500
Nitrates	45	0	45
Nitrites		0	0.1
Ammonium		<0.5	<0.5
Lead	0.5	0	0.4
Organic Chlorine Compounds		0	0.025
Pesticides		0	0
Iron	0.3	<0.2	0.2
Manganese	0.05	<0.05	0.05
Copper	0.05	<0.05	0.1
Zinc	5.0	<0.1	0.1

CHECKLIST 7: SEWAGE DISPOSAL			
1.0	**Proposed Disposal method:**		
	a. Municipal System:		
	b. On-site treatment/distance:		
	c. Off-site disposal/distance:		
	d. Sludge off-site/distance:		
	e. Filter Bed/Percolation/distance:		
	i. Size of filter bed/distance:		
	ii. Is percolation data available:	Yes	No
2.0	**Proposed effluent discharge into:**		
	a.Ocean/Sea:		
	b.Lake:		
	c.River:		
3.0	**Size of municipal main:**		
4.0	**Distance to municipal main:**		
5.0	**Elevation of municipal main:**		
6.0	**Present capacity factor:**		
7.0	**Adequacy over next 5 year**		
8.0	**Reliability of system:**		

CHECKLIST 8: FIRE PROTECTION			
The following are services normally provided by municipal authorities. Please identify each service available and furnish as much data on each as possible:			
1.0	Identify Fire Department having jurisdiction		
2.0	Emergency responders estimated response time:		
3.0	How is alarm transmitted:		
4.0	Type of firefighting equipment:		
5.0	Describe rescue equipment:		
6.0	Type of water connection(s) into building:		
7.0	Volume of required water reserve:		
8.0	Does a fire code exist:	Yes	No
	If yes, what standards are acceptable (NFPA/BS, etc.):		
9.0	Is copy attached	Yes	No
10.0	Inspection services during construction:	Yes	No
	Final Inspection:	Yes	No
11.0	Approvals during construction:	Yes	No
	Final Inspection:	Yes	No
12.0	Does controlling agency issue certificate of compliance:	Yes	No
Remarks:			

CHECKLIST 9: BOILER FUEL OPTIONS			
1.0	Identify each fuel - type available and cost:		
	a. Gas	$	
	b. Fuel Oil	$	
2.0	Bulk delivery time for each type:	$	
	a. Gas		
	b. Fuel Oil		
3.0	Are roads accessible year round:	Yes	No
4.0	Largest on-site storage permitted:		
5.0	Do emissions standards exist (attach):	Yes	No
6.0	Identify applicable codes:		
Remarks:			

Appendix 1 – Acoustical Performance

(a) **General**

The overall objective of the performance standards in this section is to provide acoustic conditions for Multi-Use buildings that:

- Allow the highest quality office spaces and retail outlets for office staff, office staff and visiting customers
- Provide a feeling of world class accommodation

Performance standards on the following topics are specified in this section to achieve this objective:

- Internal ambient noise levels
- Airborne sound insulation between spaces (that is, walls and floors)
- Impact sound insulation of floors
- Reverberation times in circulation spaces

(b) **Pre-development Noise Surveys**

Prior to commencement of the project, a minimum of 24 hours continuous monitoring shall be undertaken at a number of locations around the development site.

Data must be collected to enable:

I. Significant existing noise and vibration sources to be identified

II. Noise levels incident on each façade of the hotel development throughout the day to be determined

III. Resultant vibration levels within the development to be estimated

External Noise Intrusion

Location	External Noise	Ser vices Noise
Public spaces	40 dB LAeq	NR 35
Residential Apartment – night-time	27 dB LAeq 45 dB LAmax	NR22
Residential Apartment – daytime	32 dB LAeq	NR25
Meeting Facilities – daytime	35 dB LAeq 50 dB LAmax	NR30
Spa and Fitness Centre – daytime	40 dB LAeq	NR40

External building fabric, including any ventilators, must be designed and built to ensure that the following internal noise levels are not exceeded due to any regularly occurring external noise source:

- Public spaces are considered to include public restrooms, bar and restaurant corridors. Daytime relates to 0700- 2300 hours and night-time to 2300-0700. Unless otherwise specified, all noise criteria relate to L_{eq} noise units.

- Noise from external sources is to be measured over a 30-minue period (note that, for either daytime or night-time, the maximum noise level must be met during any 30-minute period). Noise measurements of external noise break-in shall be carried out 2m from the external wall and 1.5m above floor level. It should be noted that noise from external sources may include noise arising from operations within the hotel itself (for example, a noisy kitchen).
- Measurements of building services noise shall be carried out 1.50m above floor level and at a distance of 1.5m from the nearest diffuser or noise-radiating surface.
- Where the services noise is steady, it must be sufficient to measure the level until a steady reading in dBA is obtained (this should be not be less than 30s in duration
- Equipment installations must be designed and installed to ensure that atmospheric noise does not result in the above internal values being exceeded.

(c) **Internal Sound Insulation**

Sound insulation between vertically and horizontally adjacent spaces, including crosstalk via ductwork and service risers, should achieve the minimum levels of performance set out in above table.

Where rooms require enhanced acoustic privacy from entrance lobbies or corridors,doors and frames shall be of a design that has been proven by laboratory testing (with full supporting documentation being made available to Owner to achieve the following minimum performance standards.

Space Description	Weighted Sound Reduction Index of Doors, STC
Guestroom entrance	32dB
within 5m of lift/lift lobby)	35dB
Meeting/function room entrance	35dB
Offices	30dB
Shopping Mall	40dB

(d) **Reverberation Time**

- The reverberation time in meeting rooms and leisure facilities shall not exceed 1 second in any single octave band in the frequency range 500 to 4,000 Hz.
- The reverberation time in corridors serving guest bedrooms shall not exceed 0.7 seconds in any single octave band in the frequency range 500 to 4,000 Hz.
- The reverberation time within atria shall not exceed 1.7 seconds in any single octave band in the frequency range 500 to 4,000 Hz.

(e) Vibration from all building services plant shall be imperceptible within public areas within the hotel. Vibration from mechanical services plant shall be deemed to be imperceptible if it does not exceed Curve 1 of BS 6472-1: 200

Appendix 1I – Criteria Matrix

SAMPLE CRITERIA MATRIX								
OFFICE BLOCK RELATIONSHIP	SQUARE METER NEEDS	ADJACENCIES	PUBLIC ACCESS	DAYLIGHT AND / OR VIEW	PRIVACY	PLUMBING	SPECIAL EQUIPMENT	SPECIAL CONSIDERATIONS
1 - LOBBY RECEPTION								
2 - OFFICE ADMIN.								
3 - CONFERENCE ROOMS								
4 - OPEN WORKSPACE								
5 - TEAM SPACE								
6 - CUBICLES								
7 - PRIVATE OFFICES								
8 - SHARED OFFICE								
9 - TEAM ROOM								
10 - STORAGE SPACE								
11 - PANTRY								
12 - BREAK ROOM								

SECTION 17.0

Glossary

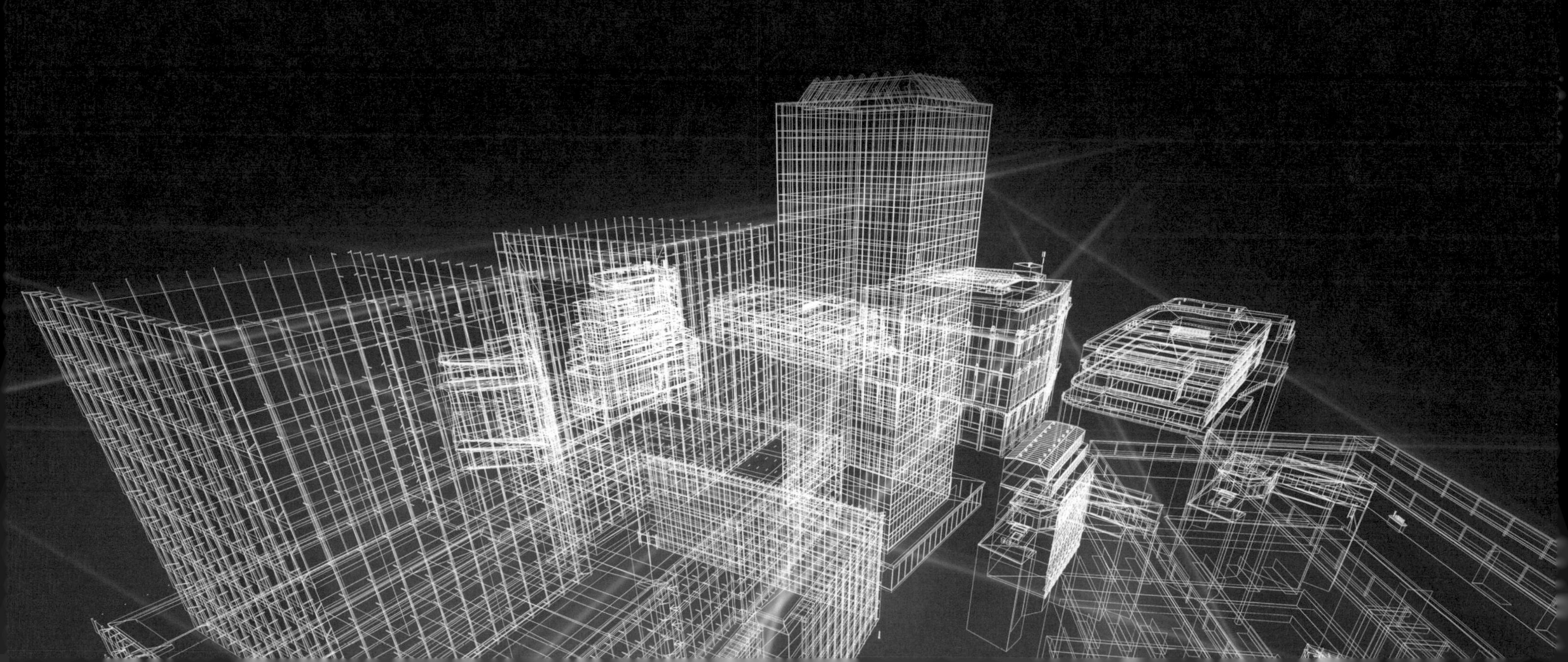

Acronym	Definition
ASTM	American Society for Testing and Materials
ACI	American Concrete Institute
AEI	AEI Music Network
AFF	Above Finished Floor
ADA	Americans with Disabilities Act
ADAG	Americans with Disabilities Act Guidelines
ADSL	Asymmetrical Digital Subscriber Line
ANSI	American National Standards Institute
AHU	Air Handling Unit
AC/DC	Alternating Current/Direct Current
AC	Air Changes
AATCC	American Association of Textile Colourist and Chemist
AP	Access Point
AV	Audio Visual
ANPR	Automated Number Plate Recognition
ASHRAE	American Society of Heating, Refrigerating, and Air- Conditioning Engineers
BMS	Building Management System
BIOS	Basic Input/ Output System
BGM	Background Music
BS	British Standards
CMU	Concrete Masonry Units
CC	Cold Cathode
CCC	Commodity Credit Corporation
CRI	Carpet Rug Institute
CPU	Central Processing Unit
CD ROM	Compact Disc-Read-Only
CAD	Computer Aided Design
CAT	Category
CV	Calorific Value
CAT	Category e.g. CAT6 cabling
CATV	Community Antenna Television

Acronym	Definition
CCTV	Closed Circuit Television
CO2	Carbon Dioxide
CPSC	Consumer Products Safety Commission
CPE	Customer Premise Equipment
CNC	Computer Numerical Control
CFC	Chlorofluorocarbon
CIBSE	Chartered Institution of Building Services Engineers
DES	Data Encryption Standard
DIN	Deutsches Institut für Normung (German Institute for Standardistion)
DDA	Disability Discrimination Act
dB	Decibel
dBA	A-weighted Decibels
DB	Dry Bulb
DHWS	Domestic Hot Water System
DID/DDI	Direct Inward Dialling/Direct Dialling In
DOS	Denial of Service
DAC	Digital to Analog Converter
DMX	DMX Music
DSL	Digital Subscriber Line
DVD	Digital Versatile Disc
ESC	Energy Supply Company
EBITDA	Earnings Before Interest, Taxes, Depreciation and Amortization
EN	European Standard
EPO	Emergency Power Off
FM	Frequency Modulation
F	Fluorescent
FF&E	Furniture, Fixtures & Equipment
FSTC	Field Sound Transmission Class
FCU	Fan Coil Unit
G bps	Gigabits per second
GHz	Giga Hertz

Acronym	Definition
GFI	Ground Fault Interrupter
GFCI	Ground Fault Circuit Interrupter
GRMS	Guestroom Management System
GSM	Grams Per Square Meter
HVAC	Heating Ventilation and Air Conditioning
HAC	High Alumina Cement
HV	High Voltage
HID	High Intensity Discharge
HSIA	High Speed Internet Access
HVM	Hospitality Voice Messaging
HSE	Health and Safety Executive
HTNG	Hospitality Technology Next Generation
HDMI	High-Definition Multimedia Interface
Hz	Hertz
ICF	Insulated Concrete Form
IDS	Intrusion Detection System
IEEE	Institute for Electrical and Electronic Engineers
IR	Infrared Receiver
I/O	Input/ Output
IP	Internet Protocol
IP PBX	Internet Protocol Private Branch Exchange
IP TV	Internet Protocol Television
IDF	Intermediate Distribution Frame
IP Rating	Ingress Protection Rating
ISDN	Integrated Services Digital Network
IT	Information Technology
IBC	International Building Code
IPS	Intrusion Protection System
ISP	Internet Service Provider
ISSU	In Service Software Upgrade

Acronym	Definition
KW	Kilo Watts
Kwh	Kilo Watt Hour
KVAr	Kilovolt-Ampere Reactive
LAN	Local Area Network
LCD	Liquid Crystal Display
LED	Light Emitting Diode
LV	Low Voltage
LPG	Liquid Petroleum Gas
LSZH	Low-Smoke-Zero-Halogen
LAeq	A-weighted Equivalent Sound Pressure Level in dB
MATV	Master Antenna Television
MIMO	Multiple Input Multiple Output
MCU	Multiport Control Unit
MDF	Main Distribution Frame
MEP	Mechanical Electrical Plumbing
mps	Meters Per Second
mpm	Meters Per Minute
MV	Medium Voltage
NAC	Network Access Control
NFPA	National Fire Protection Association
NIC	Network Interface Card
NSPI	National Spa and Pool Institute
NR	Noise Rating
NC	Not Controlled
NIC	Network Interface Card
NEC	National Electrical Code
OSHA	Occupational Safety and Health Administration
OITC	Outside-Indoor Transmission Class
o.c	On Centres
PAP	Password Authentication Protocol
PSU	Power Supply Unit

Acronym	Definition
PTT	Public Telephone & Telegraph
PoE	Power Over Ethernet
PC	Personal Computer
POS	Point of Sales
PH	Power of Hydrogen
POTS	Plain Old Telephone Service
PVC	Polyvinyl Chloride
PBX	Private Branch Exchange
PCB	Polychlorinated Biphenyls
PIR	Passive Infra-Red
PMS	Property Management System
QOS	Quality of Service
RFID	Radio-Frequency Identification
RF	Radio Frequency
RCD	Residual Current Device
RH	Relative Humidity
SPL	Sound Pressure Level
SD	Standard Definition
STC	Sound Transmission Class
SC	Subscriber Connector
SNMP	Simple Network Management Protocol
STP	Spanning Tree Protocols
SSH	Secure Shell
SLA	Service Level Agreement
SIP	Session Initiation Protocol
TACACS	Terminal Access Controller Access Control Systems
TSDG	Technical Standards and Design Guidelines
TV	Television
T	Tubular
TCP/IP	Transmission Control Protocol/ Internet Protocol
UV	Ultra Violet

Acronym	Definition
UPS	Uninterrupted Power Supply
UL	Underwriters Laboratories
USB	Universal Serial Bus
UTP	Unshielded Twisted Pair
VGA	Video Graphics Array
VOC	Volatile Organic Compounds
VSD	Variable Speed Drive
VRF	Variable Refrigerant Flow
VRV	Variable Refrigerant Volume
VA	Volts X Amperes
VAV	Variable Air Volume
WAP	Wireless Access Point
VLAN	Virtual Local Area Network
VPN	Virtual Private Network
WAN	Wide Area Network
WiFi	Wireless Networking Technology 802.11 and above (Wi-Fi Alliance)
WAN	Wide Area Network
WAP	Wireless Access Point
WEP	Wired Equivalent Privacy
WLAN	Wireless Local Area Network
WHO	World Health Organization
WPA	Wi-Fi Protected Access
WC	Water Closet

Printed in the United States
By Bookmasters